To Mary,
Dreams are the
life's daily
Keep the

15/10/98.

Making It Happen

Scribe Publications

Maureen Walsh-Martin, who was born in 1942, farmed with her husband Bernard in the Victorian Mallee for 27 years. Over these years she developed a passionate involvement in rural issues and a career dedicated to improving the access to services of rural women.

Between 1986 and 1990, Ms Walsh-Martin worked to help Mallee farming families overcome the effects of the rural downturn by co-founding and co-ordinating a farm lobby group; by working with the Mallee Women In Agriculture group in assisting 900 farm women to receive unemployment benefits; and by establishing a regional information and resource centre, and setting up numerous education programs. In 1990, Maureen Walsh-Martin was awarded the Order of Australia medal for her work with Mallee farming families.

In 1990, following the death of her husband, the Office of the Status of Women employed her as the consultant for rural and isolated Australia as part of their national domestic violence program. Upon moving to Bendigo in 1991 she was employed by a farming group to organise activities, programs, and discussion groups for farm women throughout north-central Victoria.

In addition to working on *Making It Happen*, her first book, Ms Walsh-Martin has been managing a business she owns in partnership, and is currently a member of the North Central Women in Agriculture Group and the consultative committee for the Rural Women's Network. Maureen Walsh-Martin has remarried, and lives with her family in Bendigo.

To Jenni Mitchell and Heather O'Connor, two women whose vision and work for Victorian rural women in the 1980s has been instrumental in putting into place many of the networks and resources available today.

And to my late husband Bernie and present husband Ron, both of whom have always given me their total support and encouragement to step out and make 'it' happen no matter what 'it' was or is.

Making it Happen

How 25 rural women turned ideas into businesses

EDITED BY MAUREEN WALSH-MARTIN

SCRIBE PUBLICATIONS
MELBOURNE

Scribe Publications Pty Ltd
PO Box 287
Carlton North, Victoria 3054
Email: scripub@ozemail.com.au

First published 1998 by Scribe Publications

The interviews on which this book is based were conducted by Maureen Walsh-Martin for the Foundation for Australian Agricultural Women with the help of a grant from the Reichstein Foundation

Interviews transcribed and edited by Myfanwy Jones
assisted by Sandon McLeod
Cover design by Pauline McClenahan, Captured Concepts
Cover colour photographs supplied by Tourism Victoria

Typeset in Adobe Garamond and Futura Condensed by the publisher
Printed and bound in Australia by Brown Prior Anderson

National Library of Australia
Cataloguing-in-Publication data

Making it happen: how 25 rural women turned ideas
into businesses.

ISBN 0 908011 35 0

1. Women in business - Australia - Case studies. 2. Rural women - Australia - Case studies. 3. Self-employed women - Australia - Case studies. l. Walsh-Martin, Maureen Anne.

338.04082

Royalties from the sale of this book received by the Foundation for Australian Agricultural Women will be distributed to disadvantaged farm families

Contents

List of information boxes

Foreword

Women have been farmers since the human race began to settle and turn to agriculture. In the developing world, women are still the major primary producers of food and livestock, and they play an essential role in their families' physical survival. In Australia, women are filling the agricultural colleges, and are taking their place in increasing numbers beside their husbands, brothers, and fathers as farmers. They often take up farming while raising families, managing the household, and working as volunteers in their communities.

Rural women are remarkably resourceful, capable, and courageous when faced with a loss of income or the possibility of losing their properties, often on account of factors beyond their control such as bad seasons and low prices. In such demanding situations, many of them turn their minds and hands to all kinds of imaginative and productive projects. This book tells the stories of some of these women in their own words. It tells us, in vivid detail, what they did and how they did it.

I have always had the greatest respect and admiration for our rural communities. I am very proud that the qualities of our pioneering women have been passed on, and have inspired and strengthened those who have achieved so much to sustain our rural industries.

The Foundation for Australian Agricultural Women was established in 1994. This book was initiated by the foundation to encourage rural women, particularly those struggling with survival on family farms. I am very happy to recommend *Making It Happen* as a most interesting and heart-warming tribute to them.

Dame Elisabeth Murdoch AC DBE
Cruden Farm

Preface

When I was growing up, I used to spend a great deal of time with my grandmother. She had spent most of her married life living on a farm, and she had an exercise book in which she kept all her financial records. She would often tell me of her and her family's struggles to survive over the years, and I especially remember her telling me how they would have lost the farm and often would not have had food if it hadn't been for their cow and chooks. In telling me, she would show me the records in the exercise book detailing when she had sold a calf, milk, butter, or eggs. As a result, I always thought of the cow as a truly wonderful animal which would produce a calf to be sold just in time to meet a looming bank payment.

In my early teens we moved from a provincial town back to a farming area where my mother had been raised. It was here that I had my first real experience of farming life. I would spend my weekends at a friend's house on the farm, and one of the chores was to collect the eggs daily. This was not just a matter of going to a 'chook shed' and picking up the eggs; it entailed looking in haysheds, old machinery, and various other spots where the hens liked to lay. They were truly free-range chooks. The eggs were packed in boxes and taken to town to the 'egg floor', where my father worked, to be graded and sent to the nearest regional centre. The cheque was then sent to my friend's mother. She, too, spoke of the egg money as an income to be valued.

My appreciation of the value of selling farm produce — eggs, in particular — was reinforced by my father speaking admiringly about the farm women bringing in their boxes of eggs to the 'floor'.

In my late teens I met my future husband. He was a farmer in the Mallee and, of course, I spent a great deal of time at his parents' house on the farm. Some of the chores I helped my mother-in-law with included collecting the eggs, getting in the house cows for milking, and helping with the separator. The milk and eggs were put in 'billies' [billycans] to be taken into town to the few people she sold them to. The remaining eggs were packed in boxes to be taken to the station and sent by train to the 'egg floor' which was 50 miles away. My mother-in-law also raised, killed, and dressed turkeys for sale. She, too, would tell me stories of the times during the 1930s and 1940s when she and her husband were only able to meet the bank payments because of the money received from the sale of the cows and calves she had reared.

By the time we were married in the early 1960s, regulations and restrictions had been brought in which led to the sale of these goods being quite limited. Milk could no longer be sold in billies, therefore a house cow became a rarity. And as farmers were no longer allowed to keep more than 30 hens without being registered, and all eggs had to be sold through the egg board, poultry was only kept for our own use.

Luckily for some, including me, this period coincided with the changing political and social climate which meant that it became acceptable (and possible, in the case of teachers) for married women to return to the paid work-force.

As I soon discovered, the reality of farming is that, no matter how well farmers budget and plan, they are still at the mercy of circumstances outside their control. Factors such as adverse weather, fluctuating markets, and low commodity prices can significantly reduce or entirely negate farm income for months — or, as we have seen in recent times, for several years.

At these times, farm families are forced either to use up their reserves or borrow from lending institutions just to maintain their farming enterprises. With most if not all financial resources having to be used to maintain the farming business, farm families have to find a way to continue to pay for the necessities of life, especially when as a general rule they do not qualify for any income support.

The solution often involves at least one of the men in the enterprise either going to regional towns or cities to gain employment, or finding work within farming-related industries — such as shearing, milking, and truck- or tractor-driving.

In the 1980s, a standard joke was that a viable farmer was one who was married to a nurse or a teacher. This was not a joke to the many farm women employed in these professions who were often travelling long distances in order to obtain work, or to their partners who were still working long hours on the farm for little or no income.

But what of the farm women who are not teachers or nurses? How can they contribute to the farm coffers? The opportunities for paid employment in rural areas with high unemployment and few job vacancies are fairly negligible. The answer, for many, has been to set up a business which can be either run on or from the farm. Setting up your own business can be a daunting task for anyone, but even more so for someone living in areas isolated from advice, assistance, services, resources and, especially, markets.

Over the years of living in and working for farming communities, I would quite often hear comments along the lines of, 'Farm women nowadays are not like they were in my mother's or grandmother's day when they milked a cow, kept poultry, and sold the butter, milk, and the eggs'. This statement used to frustrate me, because I knew from my contact with farm women and their families that many were working to subsidise the farms in a different way from the old days.

I mentioned this to the co-ordinator of a farming organisation I was working for. He told me that, if I believed this to be the case, I should find out and write it up. I thought he was right, so I set out a proposal for funding to write such a book as part of my paid work. Unfortunately the farming organisation ceased to function, and I moved on to managing a business that I owned in partnership.

However, Helen Morris from the Foundation for Australian Agricultural Women, who had been shown my proposal, contacted me. Helen said that they liked the idea, and asked if I would be willing to continue with the project. I agreed to do this during any free weekends I could snatch from our business. I did not then know what a long trip I was setting out on.

The first step was to find women on the land who would be willing to meet with me and have their stories put on tape, to be transcribed and edited at a later date, so that their interviews could be published in a book.

It was not a difficult task to find women to interview, as the networks both formal and informal in rural Victoria, which were established during the 1980s, are quite strong. I contacted the majority of the women by telephone, and all of them agreed to take part. I am indebted to these women, most of whom gave up valuable time with their families at the weekends to fit in with my schedule. I am also aware, having spent the major part of my life in a rural community where anonymity is impossible, that it is not an easy task to have your struggles and successes made available for open scrutiny.

I set out originally to speak to women who operated from what we would look on as farms — that is, long-established family farms. I was also interested in the many women on small acreages, commonly termed hobby farms, were using their initiative and skills to set up enterprises which were assisting in maintaining the land as well as bringing in an income. I chose not to interview any of the growing numbers of women in full management of large agricultural properties — their stories would fill another book entirely.

Upon arriving at a farm house, it never ceased to amaze me to find a thriving enterprise established in an area which seemed quite remote from even a major regional town, let alone the business world at large.

In the end, I gathered the stories of 25 rural women from a variety of farm industries, situated on both small and large acreages. All of them are using their skills and determination to overcome the barrier of distance from markets and their isolation from services to turn their ideas into money-making enterprises. The annual turnover of the businesses ranges from under $2,000 to over one million dollars.

The reasons the women began these ventures are as varied as the businesses themselves. They range widely — from a desire to diversify their farms' products as a hedge against harsh global trends to a wish to provide their families with little luxuries that they wouldn't normally have. Most women, especially those on the land, would identify with the need for extra income to:

- Compensate for the downturn in commodity prices
- Put into a specific landcare interest;
- Provide for their children's education;
- Pay off a farm mortgage or undertake home improvements;
- Establish a new interest when the children have left the farm to move to the city; and
- Assist in the promotion of their industry and community

Over the 18 months I spent travelling around the countryside interviewing these women, I realised that, while my original idea had been proved right, I had under-estimated the range and scale of the contribution rural women such as these were making. I could also see that their stories would provide inspiration and encouragement to other rural women who might be considering taking the next step of turning a skill or hobby into a business venture. To help with this, the book is interspersed with information boxes dealing with resources, associations, facts, and departments which I hope will be of considerable assistance for anyone who wishes to 'branch out' but isn't too sure about where to begin or who to contact. We have also included an extended chapter at the back of the book containing further information relevant to many of the issues raised in the interviews.

Another important aspect of survival on the land, which comes through in the majority of stories in this collection, is the way in which the skills of all family members are used in the practical implementation of the business ideas which were begun by the women. There is a common theme, running through most of the stories, of the male farmers using their ingenuity and practical skills to adapt or make suitable equipment for the new businesses. Also, there is constant mention of the work done by offspring of all ages to help make the businesses a success.

One participant in the book remarked, 'Because my business is run from the same base as the farming business, we are reliant on family and friends to come out here to check on things when we are away either on business or for a break. Without all these people this business could not happen.'

However, whatever support, assistance and encouragement were given, the idea and the drive began with the farm women. It was they who thought of an idea, and then had the courage and persistence to make it happen.

I hope that this book will not only encourage others who may be thinking of setting up their own business to do so, but may also in some small way go towards acknowledging the women on the land who are contributing so much to the maintenance of the land, their communities, and the economy. These women, in my mind, are producing the butter, milk, and eggs of today.

Acknowledgements

I have always been a compulsive reader, but must admit that the acknowledgements page has never been my top priority when picking up a book to read. Since being involved in this book my attitude has changed, as this page is the only recognition given to the people without whom this publication would not have happened.

I wish to firstly thank again the women who were interviewed for the generous giving of their time and the sharing of their lives and business details. I appreciate their patience in being available so many times to answer my follow-up queries, and I also want to thank their families who were so welcoming when we arrived at their homes, generally at weekends.

I wish to give particular thanks to Helen Morris from the Foundation for Aaustralian Agricultural Women, without whose vision and encouragement and trust in my ability this book would not have happened. My thanks also go to Helen for the huge number of hours she put into this project, not only in assisting via her networks in locating potential participants, and her research and contributions to the information boxes in this book, but also for always being on the end of the telephone for me. I wish to express my gratitude to the Reichstein Foundation for its generous financial contribution to this project and for its continual financial support of the rural sector.

I wish also to thank the local government economic development officers; the staff from the Office of Rural Affairs in Horsham; Sharon Guerra, the rural financial counsellor for the Southern Mallee Region; and Mary Salce and Jennifer North. All of these people provided me with names and contact numbers of potential participants.

I owe a very special thanks to Myfanwy Jones, not only for her expert work in transcribing and editing the interviews but also for her commitment to and belief in the book, and for her continual support and encouragement. I want to thank photographer Robert Lawson, as well, who gave up many of his weekends and travelled many kilometres in order to take the high-quality photographs contained in this book. My special thanks also go to my husband Ron — for giving up his weekends and days off to chauffeur me around to the interviews, for his belief in this work and, in particular, for his emotional and financial support, especially over the past six months.

Finally I wish to express my sincere gratitude to Henry Rosenbloom of Scribe Publications for his patience and guidance, and for all the time and effort he has put into making the finished product one to be proud of.

1 Helen Almond
Jam & preserve making

An excess of fruit from the orchard, combined with compliments from sweet-toothed friends, convinced Helen Almond to have a go at making her jams profitable. In addition to her preserves, Helen now sells fruit muffins at local markets. She hopes to find a Melbourne outlet for her products.

It was the orchard that started it. The orchard was here when we moved to the property; it must be around 50 years old. I had the fruit and I thought, what do I do with the excess? We also needed the extra income. So I started making preserves and jams. I've been registered for two-and-a-half years, but I've been making it for as long as we've been here — fifteen years.

'It's now or never'

I wanted to be home-based and I haven't any other training. I'd sold my jam to friends previously and they had all liked it. Then my sister sent me a clipping out of the paper a couple of years ago about this lady who had a similar idea. I thought, well, it's now or never. Obviously I wanted to make some money out of it. Eventually.

I started selling at school fetes and markets before I was registered. I was selling the fruit as well at that stage, and then the jams started, and people started asking me to make more for them. Then a small shop rang and said they'd heard how well I presented preserves, and would I stock their shop. So then I took the next step, which was to become registered. A health inspector came around and told me what alterations I had to make to the kitchen. It wasn't difficult.

I've financed the whole thing by myself. I had some money put aside and that was what I used. I didn't get any government funding or support or anything. You virtually have to be unemployed for that; because Andrew [Helen's husband] is employed they would see this as a sideline.

We lost our irrigation water in 1995 when it was decided to phase-out irrigation within the city limits. Our trees are dying, but a lot of people know that I do the jams and, if they've got excess fruit on their trees, they'll offer it to me in exchange for some jam. Often there are trees in people's backyards and the fruit just isn't being used; I'll come home and find a container of fruit at the back door.

We're growing our own raspberries; just a small experimental plot. We use a drip system with town water, so we can regulate the water pretty carefully. The most expensive aspect with the raspberries is that they need a wind-break around them. We're experimenting with different kinds of wind-breaks to find one that will actually stay up. A natural shrub wind-break would be great but then we'd have to be watering those as well. The cost of water is prohibitive.

'I did a short course'

I did a short course on small business at the Horsham Education Centre in 1994, but the fellow that was running it seemed very negative. He virtually said, 'Just because you make a nice jar of jam doesn't mean you can run a small business'. I thought, ha! He was actually very accurate. It's hard work.

I also did a short course on jam-making in Horsham in 1996, organised by the William Angliss College. The people there were quite interesting, but they were very much into adding pectin, and they had very technical equipment. It was interesting, but there was no way you could afford the equipment and you don't really need most of it. All the women in the class could tell when the jam was ready just by testing a sample on a cold saucer.

I think you can taste the synthetic additives in commercial jams. You can put less fruit in your jams if you use additives. It gets down to 25-30 per cent fruit instead of 50 per cent. There's probably only two varieties that I make that are difficult to set, but it depends on the season, too. It's years of learning how to judge. What I make is home-made jam, so no two batches are exactly the same.

At the moment, I would say 90 per cent of the product is locally produced. Winter-time I normally just make marmalade. So you have seasonal blocks of the type of preserve that you do, and that carries you through.

I've thought about advertising, but at this stage I have enough contacts. It would be good, though, to get a bigger market outside Horsham, because there are a lot of people here who still make their own jams. I've one outlet in Halls Gap where they go very well because they've got the tourism there, and city people like home-made jams.

'The muffins keep the jams going'

I also make muffins, but I only sell those at markets. The muffins keep the jams going. Often they'll sell better than the jams. I make about 200 muffins for each market, and Mum helps me. Before each market we have a muffin day. I cook them all here in the kitchen oven. It gets very intense. You don't want to come around on muffin-making day.

When I did the William Angliss course, I met a girl there and we decided we'd do a stall together for one of the Jazz in the Park events. We hadn't looked into it carefully enough, though, and it was a bit of a disaster. The gourmet food stalls were too far away from the crowd, and the people in charge of advertising — who were quite separate to the people organising the stalls — had advertised the day as 'bring

Making Your Own Food Products

While there are state regulations, it is best to contact the health officer at your local government office, as different guidelines apply depending on the type of food involved. The officer can come to your home and advise on the requirements. The health officer or the department of agriculture should also be able to give you information about where you can have food products tested.

your own picnic'. We thought we were going to lose quite a bit of money but we did get some reimbursement.

I do the Halls Gap market regularly, and there's one out of Horsham where I have quite a good clientele. I did a large one, the Springfest in Ballarat, but the weather was bad and there was quite a large stall fee — $50 for the day. It was pretty disastrous for me.

I love doing it, but I'm not really making any money at the moment; I am at a crossroads. It's safest to stay the size I am. With jam you've got a use-by date; you have to move the product within twelve months. I need the market before I expand, so it's a bit of a catch-22.

It would be good to get into a shop in Melbourne. I just don't know whether it's feasible. Because the preserves are seasonal, if someone wants to re-order and the season's finished, you can't fill it. That's the beauty of being in a shop where you sell on consignment. You can supply what you have in stock.

If I were to get the Melbourne market and expand, I'd need to employ someone. That might be hard, because I'm not easy to work with; I'm pretty fussy. Do you leave the other person in charge of making the product or do you get them to do the public relations work for you?

It's a lot of work on your own. Not only do you have to make the product; you have to drive around to make the deliveries and then stand around talking for half an hour. At least that way you get the feedback, and that's the positive part of it. That's why I like doing markets, too: you have customers returning and telling you what they think of the product. They'll ask you for particular combinations.

Then there's the labelling. I make my own labels, but they're not self-adhesive. It's hard to know whether to spend that extra to get the adhesive labels or whether to do it the hard way. And it's the bookwork as well. I'm not very computer-literate — my husband tells me I'll have to learn — so everything just goes in a box and, at the end of the year, I have to go through it all. So there's a whole range of things that I have to be. I have to be a wife and mother, too. That's probably why I don't do more markets; I just don't think I physically could.

ENTERPRISE INFORMATION

Name
Almonds Orchards

Address
PO Box 743
Horsham VIC 3402
Phone (03) 5382 0404

Type of farm
Orchard

Year established
1995

Product
Sauces, jams, chutney, relish and muffins

District
Horsham

Numbers employed
Self

Market
Horsham, Halls Gap and local markets

Annual turnover
Under $10,000

2 Sandra Ayars
Gypsum-mine management

Sandra Ayars is one of two women in Victoria to own a gypsum mining lease, and is a qualified mine manager. She has also worked as a grain sampler and grain buyer, paving the way for other women in the industry. She holds numerous heavy-machinery licences, and has worked hands on, driving and operating machinery, and loading trucks, in the gypsum business.

I grew up on a wheat and sheep farm about fifteen kilometres out of Ultima. My dad was of the old school, which meant that I was not involved in the farm at all. It wasn't until I went out with Russell, my husband, that I had a crash course in tractor driving.

After I finished high school in Swan Hill I worked for twelve months in a clothing shop in town, but that finished unexpectedly in 1985 when Russell and I married. We had used all Russell's savings to pay for our honeymoon and wedding costs, and we were planning to use my weekly wage to pay for our everyday living costs. So we were in financial trouble right from the start; we had to take any work at all to make ends meet. I cleaned houses and sold Tupperware.

'We were desperate'

Russell then got sick with glandular fever. We were desperate, so I had to take a job as what I call an 'apprentice hole digger' with the local shire for six months, digging holes and pouring cement. It was hard physical labour, but it paid the rent. It was quite a come-down for someone who had come dux of form six, although it earned me respect from our families.

Russell was one of seven children. As the family farm wasn't big enough to support everyone, he and his older brothers, Ken and Ron, worked for shares on his uncle's property. Before we married, they were all looking for alternative ways to make money. Everyone in the district knew there was gypsum on the property, but no-one realised the potential. So when aeroplanes started flying over, checking it out, we thought, if we don't do something with it, someone else will. You see, you don't have to own a property to mine on it. You just have to go through the right channels and obtain a mining lease. Which, I might add, was fairly straight-forward back then.

The mine started just before we got married in 1985. I remember Russell asking if I knew what gypsum was. He was surprised that I did, but I remembered my dad had used it one year on the farm. We discussed the idea, and decided to go for it.

'We were self-taught'

I wasn't in the partnership in the beginning. Well, I was side by side with Russell, and we discussed everything. Back then it was a continual black hole, putting in money and never seeing light at the end of the tunnel. After about four years Ken decided to leave the business, and I bought out his share. So Russell and I are equal financial partners. Ken applied for the first lease, but I continued it on from then.

We were self-taught, learning as we went along — trial and error. When we applied for the original two leases it was very quick. Then I had to apply for renewal every five years. It was a major undertaking, and took two to three years of hard work to get it through. It is getting more controlled and strict all the time. Recently, our gypsum licences have been renewed for twenty years, which is like a vote of confidence from the Mines Department (and saves on my work load).

Russell enjoyed farming life; that's all he wanted to do. Our families kept saying we should give up the mine — they felt it wasn't paying its way —but we are both determined people, and kept plodding away. We won a contract to supply washed crystal gypsum for aerial spreading, so we employed someone to run the washing plant while Russell kept farming. At the end of that year we found that the person we had employed had earned more in wages than we had from the farm.

'If we make a go of it, you'll be able to buy any farm you want'

I said to Russell, 'This is ridiculous. We should be doing it ourselves. We should give the share-farming away and get stuck into this full-time. If we make a go of it, you'll be able to buy any farm you want'. Russell was scared to give up the farm to go with something so risky. Most farmers' sons grow up with the security of knowing they're going to inherit the family farm. He thought, if he left it, all the years he'd put in would be wasted. It was a hard decision; a pivotal point.

The scary part for me was telling the family, because they kept doubting, saying we shouldn't do it. I knew if it didn't work I would be the one they would blame. But I had faith in us. I knew we'd make a go of it, because of our determination. There was no doubting that. Of course, that year turned out to be the recession we had to have. Good timing, huh!

It's never been easy. It's still not easy; just different. It's still difficult financially because, although we turn over more, we have more debt tied up in property and machinery. We have around fifteen loans to juggle. All four of Russell's brothers have on occasion worked for us, including two currently. I feel so much pressure and responsibility is on me, because there are so many people depending on us for their income — all our employees and sub-contractors. I cannot afford to slacken off, make a bad decision, or slip up.

Back at the start I wouldn't know from one week to the next whether we could pay the rent. I used to get an allowance from mum and dad when I was at school, and then I had a job, so I had my own money. When I married, that was it; we had no money at all. I remember I couldn't just go and buy a pair of jeans like I used to at the start of winter. That really struck home.

The mine is out of the way. It's 45 kilometres west of Swan Hill and in the middle of a paddock, so it doesn't have any public exposure.

We started advertising at field days with literature, displays, and samples. I feel that one-on-one advertising is the best. Dealing with farmers is slow, because everyone wants to watch their neighbour try out the gypsum first to see if it works. They'll wait for the whole season to go past, and then might think about trying it themselves.

'Our mine has the full range'

Gypsum is a soil conditioner and a cheap source of sulphur and calcium. It's not a fertiliser. It breaks down clay soils, helps water penetrate, and increases seedling emergence. It comes in different textures and colours, and our mine has the full range.

There is copi gypsum, which looks like flour. You can spread it on freshly lasered ground, water it in, and it will start working immediately. Then there is sugar, sand or seed gypsum, which looks like white table-sugar and has a slower release time.

Crystal gypsum has a slow release potential. When it comes out of the ground it looks like honeycomb and coral, and then has to be washed and screened. It's used for aerial spreading on mountain country which can't be accessed by normal means. It can even be dumped on an air strip and have inches of rain on it, and can still be handled and spread. It's also used in plaster board, and is one of the main ingredients of cement. We are the only privately owned mine in Victoria that I know of to have crystal gypsum.

In the past, natural gypsum was considered inferior to by-product gypsum, which came from the manufacture of other products. Natural things used to be frowned upon. And some of the natural miners were quite shady, fly-by-night type operators who gave the industry a bad name. Many people thought natural gypsum was just salt. We had to teach them otherwise, and to prove that we were reliable and honest.

Now it's completely changed. People seek us out. They've begun to realise that the by-product has heavy metals and other impurities that turn up as residues in their crops. Natural gypsum is now preferred. And I'm proud to say that we have played a major part in this change of opinion through our promotion. The Mount Macedon water reserve has our natural gypsum on their banks; they heard we were the best, and they sought us out.

One thing I concentrated on was expanding and diversifying into other markets, because grain growing is so seasonal, and income can vary greatly according to the weather. We needed a consistent income in order to make all the machinery repayments, and to retain the staff we had trained. Now our powder gypsum goes into potting mix, and on to market gardens, vineyards, orchards, golf and race courses all over the state, and we bag it for home gardens. The washed crystal gypsum goes on trotting tracks. We sell raw crystal gypsum for footpaths on golf courses and public roads, private driveways and tennis courts, shed floors, and stock yards. We used to have a quiet time over the winter, but now we don't. We like to use the winter for maintenance; to paint equipment, pull it apart and fix it up. But now we have to juggle to keep up with the orders as well. We couldn't afford to go on holidays at the start. Now we can, but we don't have any spare time.

'I've often been the first woman to break tradition'

I've got a mining licence, and I attended night classes to become a fully qualified mine manager. There's only one other woman with a mine manager's licence in Victoria. I find it easier working with men than women. Mining was a man's industry, but they respect me. I'm the first woman to be nominated by my peers and to be elected to the Council of the Victorian Natural Gypsum Miners Association — the only woman amongst eight men — and I am proud of that.

I've often been the first woman to break tradition and stereotypes to do things around this area that were previously done by men only. For instance, because we didn't make a profit until about four years ago, it was a battle for nine years, and we had to do seasonal work to make ends meet. So Russell did a bit of shearing and hay contracting, and I worked for the Australian Wheat Board for five harvests, testing grain proteins. I was the first woman in this area to work as a grain sampler. Following that, I worked as a weighbridge-keeper for three seasons. I became a grain agent for Cargill, and was one of the first female grain buyers. They also employed me as a shed operator.

The next three seasons I worked for the Australian Barley Board; we had shifted to Ultima to live, and I was in charge of barley testing at the local silo. Every year there was something terrible like mould and shot grain which I had to knock back. One farmer threatened to scratch my car, and another was going to bring his gun in. It was daunting, but we needed the income to live on and to subsidise the mine. Now nearly every silo has a woman grain-sampler, which is a real breakthrough.

'It was so frustrating'

We did everything ourselves, so we had fewer costs. I remember at the start we borrowed a front-end loader from a neighbour because we couldn't afford to buy one. Every time we'd borrow something it would break, and it would cost us more in the long run. It was so frustrating. Our first truck, which we could only get by borrowing the money from Russell's uncle, was a $12,000 Nissan UD. It blew up in the end, but it was cheap compared to prices now. A few years later we traded up to a second-hand truck for $40,000. The interest rate was 25 per cent. In 1989 we brought our first new vehicle, a Volvo N10 truck for $130,000. Around this time we applied for a third mining lease over 120 hectares on a neighbouring property.

Two years ago, when my daughter Amber was born, we updated to a Ford LTS Louisville that cost $180,000 on 9 per cent interest, and we've recently set up a truck and trailer on that semi for a further $90,000. My biggest nightmare is the thought of getting a phone call to say that a driver has rolled the truck. When you have around $300,000 invested in one vehicle, a driver's skill is of paramount concern. Luckily, we've only had three minor accidents in fourteen years. I am of the opinion that we should stick to one truck and work it hard and efficiently, rather than expanding and having a fleet. To me, that would only be more risk and pressures.

We had a set-back when I nearly paralysed myself in 1992 at a horse show in Bendigo. I broke my pelvis in two places, a vertebrae in my back, a collarbone, and some ribs, and crushed everything except for my left knee and ankle. I had to learn to walk again. I spent a month in hospital, a month in a wheelchair, combined with a walking frame, and a further two weeks in the walking frame after I finished with the wheelchair. We were on the brink of losing everything at that stage. It was always such a battle. I am embarrassed to say that we had to get a wheelchair from the Red Cross, and I couldn't afford to give them a donation for the hire (needless to say, I make up for that in donations now).

It's been extremely hectic in the last couple of years since I've had Amber. Her birth (by caesarean, because of the accident) coincided with things picking up financially. We now employ about ten people on a permanent basis. When we're busy we have six on Russell's team and five on mine. I used to be down the mine, hands on, driving the equipment. But now I'm stuck in the office doing bookwork, advertising, financing — that sort of thing. I'd rather be out and amongst it. I have two part-time guys in the office, and two part-time babysitters. So I have four clear days a week in the office, and a cleaning lady one day a week. It's been hard learning to delegate, because we've always done everything ourselves and we thought no-one else could do things as well as us. But I couldn't spread myself any thinner, and I have now realised that time is so scarce and valuable that I have to use it efficiently.

'I used to get a real kick out of it'

I loved driving the excavator and loading trucks. When truck drivers came along to be loaded, not knowing I was the only one there, they would ask for my husband, or at least a man to do it. You should have seen their faces when I'd say, 'I'm going to load you'. I used to get a real kick out of it. There was one guy in particular who sticks in my mind. He had a rag out, buffing up his fuel tanks and bumper bars and all the chrome. He was so proud of his rig, and when he saw

I was going to load him he was so worried. He got up on the top of the truck and started directing me. When I was finished, he literally mopped his brow and said, 'Phew, that wasn't as bad as I thought it was going to be'. It was such a laugh. They thought I was pulling their leg, telling them that a woman was going to load their trucks. I miss that.

I had to re-do all my machinery operating licences as the Mines Department (Victoria) would not recognise the DLI [National Department of Labour and Industries] licences and vice versa. I did all the written and practical tests for the front-end loader, backhoe, excavator, and forklift with my baby in attendance, which fazed the men a bit. Needless to say, I was the only woman there. I already had my licences with the extractive industries for front-end loader and excavator driving. I can also train and supervise unlicensed drivers on the mine site, and I've got my heavy articulated licence so I can drive the trucks and semi-trailers. There's no point owning them if you can't drive them.

But I do feel that it would be carrying too much responsibility for me to drive a semi to deliver our gypsum. I wouldn't be able to lift the tyres if I had a flat, and what if something mechanical went wrong? I have driven our small truck to the local silos with grain, because it's only a UHF radio call away. I have also driven our Volvo semi to Swan Hill for a routine service wearing a nice dress. All the mechanics wanted to know who delivered it — where was Russell?

'We had our first non-working holiday'

Two years ago we had our first non-working holiday since our honeymoon. We used to go to conferences and call that our holiday. But we were working, promoting and selling ourselves the whole time.

My dad once said to me, 'One door shuts and another one always opens'. That would be at the back of my mind when something went wrong — for example, when we lost a contract to a former employee who went out and started his own mine, slashing the price and taking our major client at the time with him. We thought that was terrible. But not long after that, the client went bankrupt. So, as it turned out, it was lucky for us that he wasn't still with us, leaving us with his bad debts. It has always turned out that way: something happens, and it leads to another opportunity if you look at it the right way.

'We poured every cent we ever made into the business'

We have never had any assistance from local government. And the financial institutions were unsympathetic. The bank was the worst. We poured every cent we ever made into the business, but if I had listened to the bank I would have given up and closed down ten years ago. We changed banks two years ago. We've had a great run since then, but we haven't had hard times since then, either. In the last two years, since Amber's arrival, we've been through our biggest lot of changes since we started.

Recently I organised a major conference in Swan Hill which involved a closed-door, working-industry-only field day. There was over a million dollars worth of equipment there, and on the way out to the field day I made sure that everyone had a tour of the mine. I also organised a two-day bus tour of tourist attractions, and diverse local farming enterprises that use our gypsum. This was for the Australian Fertiliser Services Association who were holding their Australian and Victorian conferences together in Swan Hill. It was huge; over 100 people. It was good for the local economy and put us on the map, but it also took a huge amount of work.

I love living in Ultima. I like the people and the feeling of safety, but I choose a bit of anonymity sometimes. Most local people don't realise how big our business is. Only about 20 per cent of our work's local; and when I say local, it's probably 100 kilometres away. People in the mining industry respect me because they know me and what I do. I was floored when a woman I have known since I've been married wanted to offer me a job packing boxes for her mail-order catalogue. She should have known that I run our business, which is much larger than hers, but she thought I didn't work.

I think a lot of people think that, because our office is in our home. I don't physically get into a car and drive to Swan Hill every day to an office. And, as I said previously, the mine is out of the way, and most people have no idea what goes on there. In one day we may load between 30 and 40 trucks, say around 750 tonnes in the peak of the season. That's as busy as a silo during harvest.

Not many married women in this area have had careers outside of the home and family, and they find it hard to accept that I'm financially independent. I often feel that some people would only be too happy to see us go under. They think they've got it hard, but I've been through that. We crawled and clawed our way out of it. I couldn't sit back and do nothing. If I want something, I think now how can I make it happen, and set out to do it.

'The boss is here'

Some farmers would not talk to me at the start because I was a woman. They wanted to talk to Russell. I'd answer the phone and they would ask to talk to the boss. When I said I was the boss they'd wonder what was going on. They would ring back later and double-check with Russell that what I had said was correct. It's changed now. Now clients are used to me answering the phone and doing interviews. We did radio interviews at Swan Hill's field day last week, and the announcer asked if Russell was here. I said, 'Oh yes, the boss is here'. He said, 'Oh you're kidding, he's not the boss, *you're* the boss'.

I could count on one hand the people over the last ten years who said positive things like 'Good on you' or 'Well done'. They might think it, but they don't say it to your face. Whereas there were hundreds of knockers, only too eager to tell you. You've really got to believe in yourself to overcome this. Other women are the worst, the most critical. Australians like to label people, to put them in boxes. It makes them feel comfortable.

Our turnover before tax has been over half a million for quite a few years now, between \$500,000 and \$1,000,000. The trick is not how much you turn over, but how much you get to keep. For about nine years we never made a profit — we didn't have to pay tax because we didn't clear anything — then suddenly it turned around and was profitable. It changed and grew so quickly, the provisional tax nearly crippled us. It was a shock, something I didn't count on, but it's settled down now that our profit is more consistent.

'We ended up buying our own farm'

An older friend in business once said to us, 'It takes ten years to establish a successful, stable business'. We thought, he's got to be kidding. Surely it wouldn't take that long? But it was nine years before the mine was profitable enough to stand on its own two feet. It was sound advice, and kept us going when it felt as if things were taking too long, going too slow. It gave us a time frame, some perspective.

We ended up buying our own farm, which was really rewarding because of what I had said to Russell in the beginning. We had made it happen; we had earned it ourselves. Having rented a house for nine years before finally buying our own home in Ultima in 1994, and then the farm, made us appreciate it far more. Mind you, the bank owns a fair piece of it, too! Russell would agree that when it comes to decisions like that, I'm the one who suggests it and does it.

I said to Russell, 'This is the year. We'll make an offer to the neighbour for his property'. We already had the mining lease on it, but to own the land is even better. Needless to say, it coincided with a drought. We planted our first crop just in time for the drought, which really set us back financially. Nothing good comes easy. We needed to be able to sell eight semi loads of triticarlie [a grain] for the farm to break even and to put the next crop in. We stripped three, so the mine had to support the farm. Talk about coming around full circle, considering how we started out.

It is very interesting to look back and see how our decisions have had repercussions on those around us. And it is very humbling and a huge responsibility when people have asked us for advice in their own business and have acted on it. Russell and I have always done things together, but he acknowledges that I've made most of the hard decisions. You just have to believe in yourself, go with your gut feeling, your instinct.

There was a quote I read once that stuck in my mind and which I run the business by: 'If you argue for your limitations, that's what they'll be; but if you argue for your dreams, as long as they're realistic, you'll achieve them'. If you think of all the reasons why you can't do something, you won't; but if there is something you want to do, you've got to work out how to make it happen. That's always been the way I've done things. You've got to make it happen, because no-one else is going to make it happen for you.

ENTERPRISE INFORMATION

Name
Polisbet Natural Gypsum

Address
PO Box 25
Ultima VIC 3544
Phone (03) 503 96320

Type of farm
Grain

Year established
1984

Product
Natural Gypsum

District
Swan Hill

Numbers employed
Self, husband and up to 13 employees

Market
Victoria, NSW and Tasmania

Annual turnover
Over $500,000

3 Val Bray
Bee keeping

Val Bray was working for the public service in Melbourne when she suggested to her husband Kevin that they move to the bush to keep bees. Not only has their lifestyle improved, the demand for their honey is now so great that they plan to double their hives by Christmas.

I was born in New South Wales and moved to St Arnaud when I was twelve. Before we moved back up here, I worked in Melbourne for both the federal and state public service in about five different departments. My husband Kevin worked in the building industry; he was a roof plumber.

Kevin was made redundant in 1990, and I could see that he was getting a bit down. We were lying in bed one night and I said, 'I know what we could do. We could sell up (because our sons had grown up by then), and we could go back to St Arnaud or round there and get some bees. That's something you know about, and it would be nice going back to the bush'. Kevin thought it was a great idea. His father was a bee-keeper in St Arnaud.

Not long after that — that would have been around Christmas 1991 — my mum got cancer and there was no hope of a cure. She died in August 1991. That put things on the back burner for a while, which of course we didn't mind because I helped care for her. Not long before she died, I found out they were going to be offering packages in the state public service. I thought, well, sometimes things happen in funny ways don't they? So I put in for the package and got one. We sold our house and came up here, bought this place, bought a few of the necessities, and started up the business. We've been here ever since.

I'd had no exposure to commercial bee-keeping until we came here, except when we went to Kevin's mum's for Christmas holidays and Kev would give his dad a hand. I must admit, when we first started, I couldn't even put the boxes up the right way. I was putting them upside down. It's not just a matter of having hives and getting honey. There's all the hive management, and the breeding up of bees and putting in the new queens. It has many facets.

You have to move the bees to follow the honey. It's a seasonal thing. You have to have a very good knowledge of the honey floras. Over the winter, a lot of people go to the Mallee because it has ground flora and the banksias. In our case, we usually hunt around for a canola crop and put them on that. If things go according to plan, the capeweed will be out, and its pollen is beneficial to the bees.

But I didn't find it that hard learning about bee-keeping, because it's such a natural thing. It's commonsense. One thing automatically follows the next. I suppose I picked it up fairly easily because it was all hands-on. I was out there, there were only the two of us, and I had to learn. So I did.

'We started off on the wrong foot'

I'd done the NEIS course in Melbourne before I came up here [for a summary of the New Enterprise Initiative Scheme, see p. 205]. Then I got in touch with Mac Booth; he was a REV [Rural Enterprise Victoria] facilitator at that time, and he arranged for us to go on the NEIS program.

We started off on the wrong foot, under-capitalised and under-resourced. Then we had the drought. There were times in the beginning when I thought, why am I doing this? There were nights when we were shifting the bees, and it was frosty, and you'd get stung because the front had come open. You'd be frozen stiff and you wouldn't even know where your feet were, and you'd think, oh God, what a nightmare.

Mac was a help back then. It was moral support. Just to be able to talk to him was really good, because he had all that experience with other people. Sometimes, when times are tough, if you know other people are battling the same things, you don't feel half as bad, do you? A problem shared is a problem halved.

We don't get stung much now. We make sure we don't. I've got what I call my 'Ronald McDonald suit', which is white overalls five sizes too big, with the veil, the hat and the gloves — 'the whole enchilada!'

Women in Australia's Agricultural Future:

THE NATIONAL PLAN

In 1996, the Australian standing committee of state and Commonwealth chief executive officers of agencies responsible for agriculture and resource management (SCARM) set up a rural women's working group to provide advice on 'best practice in supporting women who work in agriculture and resource management'. Arising out of this new national commitment, the working group undertook to conduct a national forum for women in agriculture and resource management.

At the forum, issues fundamental to the future of agricultural industries were discussed in the context of women being partners in that future. The forum contributed to a national plan for women in agriculture and resource management — to encapsulate ways of drawing on the untapped skills and knowledge of women, and to give leadership in preparing Australian agriculture for a sustainable future within a rapidly changing, competitive global market.

The national action plan is now used by SCARM agencies as a basis for reviewing and reporting on their work with women clients. It is intended to bring a broader perspective to decision-making across the agricultural and rural sector, in that the officials involved recognise that effective change is not possible without the full participation of all the people who have a stake in the future of the sector.

For further information, see Proceedings: National Forum on Women in Agriculture and Resource Management. *Department of Primary Industries and Energy, Canberra 1997*

The business has been going since July 1993, and it's all coming together now. I do all the bookwork, but we do have an accountant for the tax. Anything that's not bee-keeping, that's related to the business, I do; plus I do all the bee-keeping with Kevin. I chop the wood for the boiler, too. I'm an excellent wood-chopper!

Last season was good. This season is average — which is all you really want, isn't it? We've met our production targets for this year, and we've got a very good market.

'I just happened to be in the right place at the right time'

We supply to subsidiaries of a multinational corporation. I contacted them when I was doing market research, and then I rang back when we got going. I just happened to be in the right place at the right time, because their current supplier was getting out of the business. I sent them off a letter with our prices and all the usual information, and an order came back. I remember that was January 1994, because in our diary I've got 'Red Letter Day' written in big letters. We've been supplying them ever since.

The orders from this company have recently trebled. We're now selling to two outlets in Melbourne as well as their Sydney and Adelaide branches. The firm is moving to using the same supplier nationally, so we're also looking at the Tasmanian market. That hasn't come to pass yet, but it's been mentioned.

We've also had inquiries from other companies via recommendations from this same corporation. They've recommended us to a firm in Port Melbourne, a wholesale grocery that supplies 58 outlets. I've had a couple of phone calls from them; they're interested, so once we get production up I'll go and see them with some samples.

I think it's a combination of things that attracts them. We supply a good product and we're reliable. I like to think I'm quite good at marketing. I've got a good rapport with people. Because of positions I've held in the past, I've had a wealth of experience dealing with people from all walks of life.

Really, getting the market is relatively easy; keeping it is where the work is. But so far, so good. Nothing's gone wrong, so I'm very happy.

'Race has nothing to do with business'

To the very best of my knowledge, I'm the only Aboriginal female commercial bee-keeper in Australia. I do know that there is an Aboriginal man that keeps bees up in the Northern Territory. We're the only two I know of that do it on a commercial basis. But Aboriginal women have been gathering honey for thousands of years, so I like to think I've come back to doing something that Aboriginal women have always done. Race has nothing to do with business acumen; I think I've proved that.

It hasn't made any difference that I'm a woman in the industry. Women are accepted as contributors in whatever way they can contribute. All the people I deal with are women. All of the purchasing officers are women. I'm very comfortable with women.

There are also quite a few women who help their husbands do bee-keeping. I know Kevin's mum helped his dad, but not like I help Kevin. She had a young family, and I don't have any children to look after. Bee-keeping is such odd hours that it would be very hard to do what I do if I had children.

I could recommend bee-keeping to anyone. The lifestyle is marvellous. I get out in the bush and think, I'd never get a better office than here. It is so beautiful. I suppose maybe that's my heritage coming out, but I love and adore the bush. I could walk you through the bush here and name every tree for you, and most of the ground flora as well. I've also got a very good knowledge of the birds and other species. If I see ants on the tree — things that Kevin wouldn't even notice — I think, there's nectar there, they wouldn't be crawling around otherwise. The ecosystem all sort of pulls together. Maybe it's inherent in me as an Aboriginal person, but I have got a much better knowledge of those things.

What we would like for the future is to expand. At the moment we've only got 300 hives, but we're looking to purchase another 400-odd, and we hope to have 800 or 900 by Christmas time.

When I've got the shed done up, which I'm in the process of doing now, I'll start selling the honey on a retail basis. There's a Pyrenees Wine and Food festival which promotes local products, and I have been invited to be part of that. I've also had inquiries from the health food shop in Maryborough, and I could place the honey in the vineyards and shops. We sell the bees' wax, too. Last year we sold it to Capilanos in Maryborough.

Our long-term plan is to build up to about 1,500 hives and bring our sons here — or at least one son. As you can imagine, bee-keeping is very laborious. They've brought in fork-lifts and things like that, but you're still on your feet a lot. We're in our fifties now, and we're not getting any younger. But I must admit, I've never been so fit in my life.

ENTERPRISE INFORMATION

Name
Redbank Honey

Address
RMB 282
Redbank VIC 3478
Phone (03) 5467 7219

Type of farm
Bee-keeping

Year established
1993

Product
Honey and bees wax

District
North-central Victoria

Numbers employed
Self and husband

Market
Victoria, South Australia, and New South Wales

Annual turnover
Over $50,000

TAWNY PORT
Base 1990

4 Cordelia Breen
Gourmet food production

It started with grape jelly sold through 'the cellar door'. Now Cordelia Breen makes jellies, jams, mayonnaise, wine vinegars and apple bread in her small kitchen on the family's winery outside Halls Gap. She predicts her gourmet food products could overtake the wines as the primary source of income if she continues to expand her business.

We've been here just over sixteen years. We came from New South Wales where my husband Bernard was an industrial chemist and I was a librarian; we certainly weren't farmers. I'd always wanted to get out of the city — I'm a country girl at heart — so when Bernard was looking for a change in career we decided to give the winery a try. It was a steep learning curve, but we learned.

Our very first year here was the 1982-1983 drought. We've probably had six or seven really good seasons and another five or six that have been so-so. The rest have been quite difficult for a variety of reasons, so I've had to look at subsidising our income.

I've been making grape jelly for a long time, and selling it since 1993 when I registered the business. With the raw materials out the back, I thought, why not give it a go? Grape jelly is not something that's common on supermarket shelves, so it sold quite well to the tourists who came into the place. Then I started to make a spiced grape jelly to go with meat. Then, after a few years, I started making jams — this has traditionally been a fruit-growing area and I grow a lot of my own fruit. We also started making herb vinegars with a commercially bought vinegar and herbs I'd grown myself.

The next thing that came was mayonnaise. My background is

German, and my mother always made mayonnaise. So I took her recipe and looked into the chemistry of it to ensure the product would keep and not poison people. Bernard was able to help me out with this, but I've also had advice from people in the food industry and testing laboratories in Melbourne. Then I started selling my mayonnaise, and it's been the mainstay of the business now for a couple of years.

I grow a lot of the ingredients myself and I buy other foodstuffs from wholesalers — like the 200-litre drums of oil. I buy jars by the pallet. Having the winery helps no end because we already have those channels open.

It's only a small one-person kitchen, but it is registered. I also do lunch for winery guests, either in a picnic hamper or served on the premises, and the occasional catering job. Last weekend we catered for 150 people at the Halls Gap Film Festival!

'I didn't fit the criteria'

When I was setting up I spoke to a number of local government people but most haven't been able to do much for me. I did get some help from the Self Employment Development Scheme [SEDS]. SEDS was a combined CES [Commonwealth Employment Service] and DSS [Department of Social Security] initiative that gave me a DSS allowance while I was getting going, between 1995 and 1996.

I did try for the NEIS scheme but things went a bit wrong there. I had to put in a business plan and they approved it, but then turned around and said, 'Okay, you know what you're doing, go ahead'. Because I already had the business going I didn't fit the criteria.

One useful thing was that they told me about the Information Victoria Christmas Markets. I participated last year and I'll do it again this year. They're wonderful markets, because we can sell direct to the public and the folk there are very helpful.

Every item has a niche in the market; no two items are the same. The mayonnaise and wine vinegars sell well in the city where there are people who appreciate fine food. The fancy herb vinegars, jams and liqueur fruit go well with the tourist market up here.

People don't come to Halls Gap expecting to spend up big. But most people like to take home something just to say they've been there. This is where my product comes in very handy, because it retails at $4 to $7 and that's well within most people's budgets.

In springtime we have a lot of tourist buses up here looking at the flowers, and other tourists trickle into the winery throughout the year. We hear quite often that either our wines or some of my products are going off to Germany or to America or whatever. It's a bit of a thrill.

In Melbourne most of the products are sold via shows — like the Country Living Show, the Victorian Winemakers Exhibition and the Information Victoria Christmas Market.

Tasting is the key. If people can taste what you're selling, 90 per cent of the time they'll buy it. That's where the shows really work for me. But each show is a gamble. Even two years in a row, the same show will be quite different.

'I did need to have at least one Melbourne outlet'

People say to you all the time at shows, 'Where can I buy your product?' I'm not big enough to get into places like Safeway — I don't want to ever get that big — but I did need to have at least one Melbourne outlet.

A friend introduced me to the Myer food buyer, who then came to see my product at the Victorian Winemakers Exhibitions. It took twelve months after that to get my product onto the shelf because I had to change my labelling and presentation to suit Myer; but I thought it was important to have the product in at least one place in the city, so I went along with it.

I've had my labels designed professionally, but at the moment I'm still making them myself. Bernard and I do that in front of television at night. The time will come when I will have to get them professionally printed, too — they do look a bit amateurish, but perhaps that's part of the image.

The tourist market is more constant. I supply a couple of shops in Halls Gap with small amounts every couple of weeks. It's probably much the same in overall income as I sell in Melbourne, but it's a constant trickle up here, whereas when I go to shows in the city it's a big burst all at once.

WOMEN'S COMMITMENTS:

'Just doing what has to be done'

In 1996 the Victorian Rural Women's Network undertook research on the issues of most concern to women involved in the network. Two hundred and six women responded to the questionnaire, listing their personal, family, and community commitments. The category of unpaid work headed the list of commitments, followed by community work and paid work (which included full-time and part-time work). 'Study' included formal study and personal development. 'Community work' included work for schools, voluntary organisations, and the environment.

Commitments that covered the home, farm, and family business were seen as unpaid work. In the words of one respondent, she was a 'wife, parent, daughter caring for aged mother, daughter-in-law, partner, cook, domestic engineer, tractor driver, sheep shifter, secretary, arbitrator, business woman, *all unpaid'*. Women who were involved in farm work also took on paid work-commitments in order to support the family farm.

It was evident that the women were combining traditional roles with an exciting diversity of interests and skills, with 33 per cent indicating business as a commitment. Most of the women indicated that they juggled a diversity of roles within the home and the community. Almost three-quarters of the respondents had three or more commitments. A significant number were undertaking study, although this rated as a lower commitment than those to family, community, and paid and unpaid work.

For more information about this research, see
More Than One Way: a report on the Rural Women's Network.
Victorian Government, Melbourne 1996

'Wine vinegar is our latest venture'

Wine vinegar is our latest venture. I hadn't thought about it until quite recently when somebody said to me, 'Have a taste of this'. It turned out to be a red wine vinegar, and it tasted like nothing I'd ever tasted before. Obviously it tasted vinegary — acidic and very sharp — but looking past that was the most amazing flavour.

So we acquired a 'vinegar mother', which is a gloopy, jellyish, horrible-looking thing. It's a bacteria that works on the alcohol in wine and turns it into acetic acid — which is what vinegar is. We've just started producing Shiraz vinegar on a commercial basis and, while it's bitey like vinegar is, there is a really good fruit characteristic behind it. It's wood-aged, so it's got a touch of wood flavour there, too. It's much nicer than the red wine vinegars you can buy at the supermarket. It's surprising how delicate a vinegar can be; it sounds like a contradiction in terms.

I've had fantastic responses to the red wine vinegar. I've had it out for tastings, and people thought it was wonderful. I'm now working on a white wine vinegar which is going to have a strong, almost sweet, fruit flavour to it — getting close to a balsamic vinegar in flavour. I'm really pleased with it. I'll have that commercially available within the next few weeks.

The next thing we're going to try is a mead [honey wine] vinegar. I've got the mead fermented now, so I'll start setting that up shortly and see how it goes.

My newest product is 'apple manna'. One of my friends has an apple orchard and I've been making apple jelly for quite some time, with a bit of mint to make it different. But I wondered what to do with all the leftover pulp. Rather than throw it out, like most people do, I started using it to make apple manna, which is like an apple bread. I got the name when somebody said, '"Manna" is the bread of heaven'.

Apple manna is sold in 375 mL boxes and can be sliced up for a cheeseboard or used as a confection, because it's quite sweet. I haven't taken it to any shows yet, because I've had a few problems with the consistency, but I'm selling it here through the cellar door and I think I've got it right now.

So it seems that one thing leads to another, and it's a continually evolving process. If products like mine become too abundant in the market and sales go down, I have to turn to something else. There's always a bit of a niggle in the back of my mind: what if I can't think of something new? So far I've managed.

'There is potential for it to grow without limit'

My business has bought us a holiday most years. At this stage it certainly isn't supporting us, but it is making a difference. I'm still very small-time. The potential for the business to grow is certainly there; it all depends on how much energy I have and how much I can put into it. At times I just want to say, stuff this, I'm going to sit and do nothing for a couple a days. But there is potential for it to grow without limit.

I'm talking to a lass in Melbourne about distribution. Nothing's finalised, but if we can make it work — bigger and better — I'll probably have to build a small factory and start employing people.

If that happens, it's quite possible that my products could overtake the wines. I have to make the decision as to whether I *want* to take that next step. When I started this, I said to myself, 'It has to look after itself; I'm not borrowing money from anyone to make it work'. What we have now has been self-generated. I've always said, 'I want to run the business, I don't want the business to run me'. So each step forward has been taken very cautiously, and this next step will be the same.

ENTERPRISE INFORMATION

Name
Grampians Gourmet

Address
RMB 2072A
Stawell VIC 3380
Phone (03) 5356 4502

Type of farm
Vineyard

Year established
1993

Product
Mayonnaise, jams, jellies, vinegars and apple bread

District
Grampians

Numbers employed
Self

Market
Melbourne and local area

Annual turnover
Difficult to ascertain at present

5 Lynne Bullen
Tourist maze & café management

Lynne's business has been a long-term project while she waited eight years for the trees in her maze to grow. She also had to learn commercial cookery. The business opened on New Year's Day 1997, and is finally providing a small income while Lynne keeps planning and planting.

I was born near Eltham [15 miles from Melbourne], which in those days was fairly rural. I moved to Wandiligong with my first husband in 1980. We lived on a small farm at the end of the valley, and we had two children. After our marriage break-up I bought a house in Wandiligong (Wandi for short) township, and worked as a cleaner in the Bright Hospital. But after being out on the land I didn't like the traffic going past, and so, when a house came up off the main road I bought it.

'I said, "I think I might plant a maze" '

It seems silly, but I didn't realise when I bought the house that it came with seven acres of land. When I discovered that, we were sitting on the verandah with some friends and they asked what I would do with the land. I said, 'I think I might plant a maze.' My sister had a small private maze on her farm, and I was taken with the idea. I thought I could put a maze in for tourists. The rural village of Wandiligong has been classified by the National Trust, and so you get people coming to look at the old buildings and landscape.

The idea grew, and the café plan went with it. My partner, Adrian, liked my idea and we decided to go ahead. In March 1990 we hand-planted 850 tubes of *Cypress Lambertiana* which we bought from the

Western District. Some of the locals wondered what was going on, with the mysterious circular plantings.

In the years that followed, I planted the gardens around the maze, and Adrian built the rotunda and café, all from timber. This was during the eight years it took for the trees to grow. I still worked at the hospital and Adrian worked as a maintenance man at the local gold mine. We did each stage as we could afford it. Most of our spare time was spent working on the project. The maze and garden are planted in what was originally the 'home paddock' of the small farm, and we live in the original cottage, built in the 1870s.

All the wood we've used has been offcuts or recycled, as cheap as possible. Everything that can be recycled is recycled. We try to get as much as we can from the local area, and if not local, it has to be Australian. All the timbers are Australian-grown, and have been cut and milled by hand. Apart from the local timbers, there are timbers from North Queensland where Adrian had a permit to salvage logging residue.

During the timber salvaging, Adrian had a run in with a huge python that was blocking the track. He tried to shoo it off the road, but it ended up chasing him rather than the other way around. When Adrian thought he had gained the refuge of the truck he found that his mate had locked the doors. Fortunately he survived.

'We even used an old elm branch'

A local architect, Trevor Robinson, designed the rotunda with inspiration from the grand designs of the past. We used huge narrow-leafed peppermint bush poles, and then Adrian welded the steel apex on site, which was raised on a scaffold. We needed a crane to lower the Oregon beams into place. The floor is hand-sawn broad-leaved and narrow-leaved peppermint, the railings are red gum, and the seats were crafted from silver quandong. We even used an old elm branch that had fallen down for the curvy bits on the posts (corbels).

The restaurant building also started with structural bush poles and locally grown Oregon beams, with rails of grey box and balustrades of pin oak. The verandah floor is of various rough-sawn local eucalyptus — red stringy-bark, mountain grey gum, manna gum, and narrow-leaved peppermint. The outside cladding is of bush mill flitches of eucalypt pine and Douglas fir.

The restaurant tables are all hand-crafted from North Queensland white beech and silver quandong, local red gum, and river red gum. Adrian recently used some eighteen-foot pieces of forest red gum to make two tables for the beer garden. Adrian did all the hand-milling and building at the maze, starting with the rotunda. I knew what I wanted, and Adrian worked to my ideas. Adrian is amazing with wood. I worked out the perspective and proportion, and we made it chunky looking, which I like. I had a clear idea of how the restaurant should be. I never want to stop making it a bigger and better place.

I had to discipline myself not to allow myself to do any courses that might have side-tracked me from the main aim of only working on the project. Family holidays and socialising were put on hold, because all our energies were put into making the project. It hasn't changed much. Our busiest times are when all our friends and family are on holidays.

At first the local council wouldn't let us put up any signs for the maze. But I waded through the red tape, which was tedious for a novice, and step by step I found it was only a question of filling-in forms and following instructions. The locals have been supportive, I think because the maze offers some more employment opportunities in the area.

'The café is known for the enormous sponges'

I hadn't really done any cooking before, so, just before opening the maze, I did a short course in commercial cooking at Wangaratta TAFE. The instructors at TAFE have been very helpful. They were chefs themselves, and they still keep in contact and give me a few tips. We cater for weddings and parties, as well as afternoon teas. The café is known for the enormous sponges. People walk in and say, 'There it is; I told you they were that big'. I got the recipe from a local woman, and I just doubled it; it's become well-known. The first 'Wandi Sponge' was made in the early hours of the morning while we were preparing for our first day of opening. Friends were there helping with cleaning and getting things ready — a total experiment.

Adrian works from eight to five; I work from nine to five, and in my spare time I do the gardening. I don't mind that, because gardening is my love. It's the place I love to be. Adrian makes a lot of the interesting soups. I do everything else. We employ a couple of people as casuals when it gets busy. It's just increasing in size all the time. We haven't done much advertising, because we couldn't afford it. We just advertise with the tourist places, and have a couple of brochures we made up ourselves, but mostly it's word of mouth.

I've been finding out more about people, how they're good at different things. I get impatient that everybody doesn't work the same as I do, but I've learnt that some people are good at cleaning, others at cooking, others at public contact. I've found TAFE courses very helpful. Apart from the cooking course, I've also done 'time management' and I'll be doing 'train the trainer' soon. I need to keep learning myself so I can train my staff and manage the business efficiently.

'I used to be there till two o'clock in the morning'

The hard things have been the work-load and the stress of organising everything. It took a long time to get everything done when we first opened. I used to be there till two o'clock in the morning, and still hadn't counted the money. It's much better now, but I still find interruptions hard. The maze takes an enormous amount of work. It has to be trimmed and cleaned up. It takes about two months of the year with continually looking after that. I do all the gardening.

I'm happy now. I used to get depressed and think, it's not going anywhere. It seems so hard to make money. We're doing a bit better than breaking even, which is much better than it was. I'm rapt it's successful, but we still need to get people coming all year round. I'm working at getting people from the snowfields now. People with kids need to take a day off from skiing, and we're close to Mount Hotham, Falls Creek, and Mount Buffalo. We also tap into the school trips which come up to the snow, and we advertise in the snowfields and various places the schools would go.

That was the idea behind our theme days. Last year we did our first one. We did *Alice in Wonderland* and made the afternoon tea like the Mad Hatter's tea party, and had all the things that went with the story. They've proved very successful, again because of advertising and word of mouth. This year we're doing one on *The Addams Family.* We'll have a Morticia in the maze, and scary food, so that's a further expansion.

My philosophy is to provide an environment where everyone can indulge in the simple pleasures of life — eating, music, nature, literature, art, horticulture, and socialising — in a friendly, comfortable, and exciting atmosphere. People feel so comfortable they try to bring their dishes to the sink in the kitchen. I think the natural beauty and history of the area, together with the gardens and puzzle of the maze, create a wonderful ambience. I want everything to be spontaneous, relaxed, simple, and colourful, including the food. It has to be fresh, chunky, generous, and home-made. I want it to feel the opposite of a fast-food outlet, with everything from traditional country fare to exotic spicy dishes to bush tucker. We've used a bit of local comedy on the menu, which visitors can relate to, like the 'flyers' (which includes hang-gliders and paragliders) lunch.

'The garden has been my thing'

I began the garden in the early nineties and designed it to give expression to colour, perfume, atmosphere, and season. Wandiligong has a cold frosty winter and very hot summers. Autumn is best, with mild weather until almost June. I laid out the garden for different areas to have their own distinct character, for the rotunda to be the focal point, and for the visitor to meander through the surrounds.

It's a cottage garden filled with annuals, perennials, and over 200 heritage roses. At present there's the garden around the rotunda; the water garden where I've chosen roses for the leaves and the hips rather than the flowers; the silver birch and grey garden with all red roses; the mainly white central garden; and the colour garden which is a riot of mauve, orange, and purple in the warmer months. There's a double arbour, and roses range along the fence line.

The garden has been my thing. I have always loved gardening, so I planned and planted around the rotunda. I like circles, so most things have been done in circles. No garden is ever finished, and I'm still extending: putting in more roses and the rarer plants which flourish in this climate. Adrian has designed and is now building a water wheel 30 feet high. We're going to put it up behind the maze, and there will be more gardens to go with it.

I hope it gets so busy it will run itself so I can garden more, although it would be hard to find someone else to run the café. I've realised how important personality is. People come back because we're here and they get a friendly welcome. I love the fact people come here and enjoy themselves. That's the gorgeous thing.

ENTERPRISE INFORMATION

Name
Wandiligong Maze & Café

Address
White Star Road
(PO Box 125)
Wandiligong VIC 3744
Phone/Fax (03) 5750 1311

Year established
1997

Service
Tourism establishment & informal reception centre

District
North-eastern Victoria

Numbers employed
Five

Market
Holiday-makers and locals

Annual turnover
Around $40,000

6 Susan Campbell
Quilt making

In 1992, Susan Campbell toyed with the idea of leaving her job to take up quilting. In 1995, after her youngest daughter had graduated from university, she took the plunge: she bought a quilting machine and set to work. Susan is now booked up more than a year ahead, and her business helps subsidise the farming operations.

I was born and raised in Adelaide, and I met my husband, Kim, in a pub in Melbourne. We lived in Albury after we were married — my husband's business was based there. Then we moved to Tumbarumba because they had a lot of surveying work that needed to be done up there. We lived in Tumbarumba for twelve months and then came back to live in Albury.

We moved out to Yackandandah at the end of 1976. We have ten acres here. We've owned the land since 1977, but we lived in the township because it was easy for the children to go to school. We planted the chestnut trees in 1977; that was the first lot. The next lot were planted in 1978. Then we built the house and moved down here in 1989.

In 1978, because of a shortage of work in the survey field, Kim went and did contract work in Kenya. He was supposed to go for three months, and he ended up going for nine. During that time, I managed going to work, the farm, the house, the kids — all the kit and caboodle. Then he came home for five months, but the contractor that had given him the job in Kenya asked us to go back. So in 1979 we went over and lived there — the whole family — for two years.

Before we left, we came to an arrangement with a local builder who had a few sheep and wanted a bit of land to run them on. It was sort of a trade-off: he had the use of the land, and it was his job to care for the trees while we were away.

'The quilting machine is like a huge sewing machine'

I have always sewn, right from when I was a very small child, but I didn't become interested in patchwork until my youngest daughter left home in 1990; even though I worked full-time, I was left with a lot of time on my hands. So I went and learnt how to do patchwork. Of course, if you do patchwork, the end result is that you have to quilt it. I hand-quilted my first quilt. The second quilt that I made, immediately after the first one — I must admit I became hooked, it was like a disease — I quilted under a normal sewing machine, and I wasn't happy with the result. So I just kept on making quilt tops, and by the time the end of 1994 had arrived I had about ten to twelve quilt tops that hadn't been quilted.

I worked to put the girls through university until the end of 1994. My idea was to then retire and be a kept woman, but I was bored out of my brain after three months. I had thought, in 1992, of having a look at the quilting with the idea of finishing work, coming home, and just doing that. But I wasn't game enough to take the step back then, because I still had to find the money every week to support our two girls in Melbourne — that's where they went for their university education.

So, half-way through 1995, I went to Sydney to have a look at one of these quilting machines. There was a lady there selling them. The quilting machine is like a huge sewing machine. It's got a wide arm, and it runs over a series of rollers; so instead of moving the quilt under the needle, as you do with an ordinary sewing machine, you load the quilt over this frame and then you move the sewing machine over the fabric. The table's 14 feet wide and the machine's got an 18-inch arm.

In 1992, when I first looked into getting a machine, my idea was to buy the small machine and just quilt them for myself. But when I saw the big one, I had to have the big one. It arrived here in October 1995, and I did my first quilt for somebody else on the 29th of November, 1995. That was the first one and, from then on, it's just grown.

I had worked in offices before I started the quilting, so I knew how to set up the business. I was an accounting book-keeper. Before being an accounting book-keeper, I'd worked in legal offices; so I was aware of the procedures for registering a business name and also how to run a set of books.

'I got in the car one day'

To get myself started up, and known, I quilted up all the quilts that I had. Then I got in the car one day and drove to Melbourne with my fliers. I'd produced a flier — Kim helped me with that — which just had my name and address, telephone number, and some of the patterns that I could do at that stage. I went to several shops in Melbourne, carted my wares around, and was fortunate enough that they had a look at the work that I had done and gave me a go. I knew that I was doing the best that I could do; I just hoped that that was good enough. It was. They've been very good to me, particularly Patchwork House in Hawthorn and Primarily Patchwork of Camberwell.

It's principally been word of mouth since then. I have advertised in two of the Australian patchwork magazines. One's called *Downunder Quilt*, which is Sydney-based. It's a little magazine, but I find it's very good value. The other one is the *Australian Patchwork and Quilting* magazine, which is put out by a larger organisation called Express Publications.

I've advertised in the local quilt group I belong to in Wangaratta, the Wangaratta Centre Quilters. They hold an annual residential workshop at the Dookie Agricultural College, and my advertising there is that I do a job for free. I've also done that for the Waverley Patchworkers.

This year, a group of ladies from Waverley Patchworkers made a quilt which is being raffled to raise funds for breast cancer research,

The First International Conference for Women in Agriculture: Melbourne 1994

In the evaluation of this conference, the largest issues for respondents were: family-farm viability, networking, environmental sustainability, marketing practices, personal growth, and the need for women to be heard, acknowledged, and adequately represented. This last issue was by no means the least important, as it was seen as the means by which change needs to happen. It also connected directly to the need for women to gain confidence to be able to act for change. Other consistent and important issues were: the need to see farms as part of a larger rural/cultural community; the need to bridge the gap in understanding between farmers and city dwellers; the need not to alienate men but to work with them; and the need to gain greater understanding of global economics.

Viable farms were seen to be threatened by a myriad of interconnected factors. Seasonal conditions, low commodity prices, rising costs, and impractical or insensitive government policies were often cited as difficulties facing farmers. This situation is often exacerbated by the current decline of rural towns and communities where fewer and more expensive services such as health, education, and transport are available.

This also causes young people to leave rural communities. Not only is the family farm unable to support them but there are no local jobs to go to. Often they are not only forced to leave the district as well as the farm to find employment, and never return. The pattern of inheritance is changing dramatically, and the cost to the community of losing its young people is inestimable. It is more difficult for farmers to find local work to supplement farm incomes. Increasing isolation and fragmentation puts pressure on families and individuals. Women often bear this load.

One of the strongest messages of the conference was that women can make a difference to their own lot and to agriculture generally. But, to do this, women need to be confident, educated, and able to overcome their own fears.

For more information about the conference, see Agents for Change: farming for our future. *Evaluation of the first International Women in Agriculture Conference. Melbourne 1994* and they asked me to quilt it. When it went back to them, one of them said, 'I think that can go in the group section of Quilt Showcase'. So in it went, and it won a first prize in the group section. We're very pleased with ourselves.

Some of the intricate ones that I've done take a few days to do, but there are others where less work's required — they might be only four or five hours. The one that I'm doing today, which is an appliqué quilt, I probably won't finish until 10 o'clock tonight.

I basically work on my own, but if someone asks for binding to be done on the quilt, there's a lady down the road who makes the binding and puts it on. I pay her. So I farm a little bit out that way. Every now and again someone will come to me and say, 'I have put curtains up in the house, and I've got this fabric over, will you make a bedspread for me?' So I send it out to another woman who makes the backing up to size and I quilt it. Then it goes back to the other lady and she puts the binding on. When there is an opportunity for someone else, I make sure that I pass it around.

'I'm having to think too far ahead'

I'm booked up until November 1999. I've got a query in the car at the moment for a quilt from Western Australia. I've done one for someone in the Northern Territory. I've just sent some back to Mt Isa and Sydney. But it's mainly Victoria.

I've managed to slot some time out to take holidays. We're going on holidays in a couple of weeks, and then I'll take a couple of weeks off over Christmas. The way that they've been coming in, I'm having to think too far ahead.

My husband wants to retire in a couple of years time and he said to me, 'What can I do besides becoming your lackey?' I said, 'Well, after you've done the banking and the mail and all that sort of thing, you'll have plenty to do with the day'. He said, 'No, it won't be enough'.

We had about twenty guinea fowl walking around the place, so I said, 'Why don't you start seeing if you can breed them and create a market?' I think there's something like 20 million birds eaten in France every year. I know they've got a big population, but if 20 million guinea fowl are eaten in France very year, then surely they're something we could breed here.

We've bought a little incubator, and we turn the eggs by hand. We've put in 120 eggs, and we've actually raised about 80 of those chicks. We've only done one lot, because the guinea fowl only lay for six months of the year — a bit like a chook. They start in September and finish in March. But they're silly, stupid birds; they lay all over the paddock, and you have to go and try and find the nests before the foxes get to them.

Then there are the chestnuts. When we planted them back in 1977, people were getting up to $10 a kilo. After waiting the appropriate time of about five years before we got a small crop, of course, everybody else had planted them as well. We still sell them, though this year we haven't had a crop because of the drought.

The quilting is helping to subsidise the farming operations at the moment, but we hope the guinea fowl will eventually be the one that will take off.

ENTERPRISE INFORMATION

Name
Rowdy Flat Quilting

Address
PO Box 888
Yackandandah VIC 3749
Phone (02) 6027 1387

Type of farm
Mixed horticulture & guinea fowl

Year established
1995

Product
Quilts

District
Yackandandah

Numbers employed
Self

Market
Australia-wide; mainly Victoria

Annual turnover
$15,000-$20,000

7 Loretta Chance
Pottery

When the value of angora goats fell, sending the family farm into financial straits, Loretta Chance took her hobby — pottery — and made it into a successful business. She now works full-time in her pottery studio-cum-machinery shed, and her work is sold in galleries across four states.

When we were first married, my husband Stuart ran a very successful engineering business in Melbourne. But he became ill, about twenty years ago now, and I had to take over a lot of his business. It was during this period that we decided to move out of the city. Before we left Melbourne, while Stuart was studying angora goat husbandry, I did a TAFE pottery course for three years. I was always good at arts and crafts at school, and I wanted a hobby.

Fourteen years ago, we bought this 550-acre farm to raise angora goats. But, shortly after purchasing the property, the goats started to fall in value. My husband had to go and work off-farm for other farmers in the district, and I was left alone for the first time in my life. All the children were at school.

'I needed to find something to do'

I had all this time on my hands and needed to find something to do. After speaking to people at Horsham TAFE, I realised that, while there were other potters in the area, there was nowhere local to buy equipment or to get pottery fired. This gave me the idea of setting up a kiln and supply business in town. I envisaged that the shop would be able to both keep up my interest in pottery as well as provide us with a good extra income.

I didn't seek out help from any government bodies or other organisations while I was setting up the business because I didn't know any of it existed. After I had been running it for about two years, I heard that there was a Women's Enterprise Scheme, but because I already had the business going I didn't qualify. If I'd known about it two years previously it would have been very handy to get a grant or low-interest loan. We had to buy a slab roller and other equipment to start up.

Anyway, I soon found that people wanted to come and buy my pots rather than make them themselves; so after twelve months we closed the shop. The kids and I cleaned out an area in the machinery shed and Stuart helped us put windows in and put up a mud-brick wall. We concreted the dirt floor to keep the dust down. Then we put a door in, and the kiln and wheel, and away I went. I set to work making pottery in my little studio.

'The first gallery I ever got into'

Initially I was selling my work in Horsham and at a few craft markets around the place. That didn't last very long because I was doing very fine white-ware — porcelain platters and bowls and vases — while most of your craft-market people want mugs and that sort of stuff. I needed a more steady income.

I had contact with some potters in Ballarat, and they gave me the names of a few galleries in Melbourne they were selling their work through. I filled up the back of the station wagon with a couple of hundred pieces and off I went.

The first gallery I ever got into was Potters Cottage at Warrandyte. Most people in Melbourne, if you say anything about pottery, will say, 'I've been to Potters Cottage'. They've got 50 or 60 potters supplying them. I made an appointment with the fellow there and went in with three big baskets full of work. I was very careful about the finish; your work can't have any faults in it. Fortunately he liked my pieces and bought them straight away. That gave me the confidence to go on to the next place. Then, every time I went to a gallery and they asked me where I had my work displayed, I could say, 'Potters Cottage'. They know if you've got work there you must be good.

'The recession hit'

After I'd built up to about ten different galleries, the recession hit. People who had been willing to spend $300 on a platter for a wedding present in the late eighties were starting to look at a casserole dish for perhaps eighty dollars. A lot of potters found that time very distressing. Because I was working on my own I could turn out whatever I liked, so I turned my pottery over to more domestic ware. Previously I had been only making and selling individual, and therefore expensive, pieces; I'd never made a casserole. Other artists, like painters, were hit much harder by the recession because they couldn't just turn around and make prints.

I became so busy at that time that Stuart couldn't do any outside work. The pottery business was almost seven days a week because I was churning out lots more but not getting as much per item. We still had to keep up with our shearing and animal husbandry on the farm. We had about 800 goats, and goat prices were very low at that time; so we sold or gave away 400 of them and put some sheep in. We had a few merinos, and we cross-bred them so we had meat lambs.

Now the economy has improved again, I am back to making fine porcelain ware. I specialise in functional and decorative ware for the mid-range market — colourful, fine stuff that you can put in the dining room — but I don't make one-off pieces any more except for special exhibitions. Galleries are more particular about what they buy now, too; ten years ago people thought pottery was something brown and heavy that you kept in the back of the cupboard in the kitchen.

I have outlets in Melbourne, Adelaide, Western Australia and New South Wales, but I still get nervous when I have to approach a new gallery to show them my work. I always look at what other people are doing and wonder whether mine is good enough.

Marketing is probably the most difficult part for any small business or enterprise. Because of our previous experience, having had an

National Crafts Council Network

There is a craft council in each state, each of which has a central information service and a membership from which its management board is elected. Each council also carries out a range of other projects particular to the needs of that state. Their names and addresses are:

Crafts Council of ACT
1 Aspinal Street
Watson ACT 2602

Centre for Contemporary Craft
88 George Street
The Rocks
Sydney NSW 2000

Crafts Council of NT
GPO Box 1479
Darwin NT 0801

Crafts Council of QLD
166 Ann Street
Brisbane QLD 4000

Crafts Council of SA
PO Box 8067
Hindley Street
Adelaide SA 5000

JamFactory Contemporary Craft and Design
PO Box 10090
Gouger Street SA 5000

Crafts Council of TAS
Bellerive Community Arts College
17 Cambridge Road
Bellerive TAS 7018

Tasmanian Design Devlt Co
Hunter Island Design Centre
19A Hunter Street
Hobart TAS 7000

Craft Victoria
114 Gertrude Street
Fitzroy VIC 3065

Meat Market Craft Centre
42 Courtney Street
North Melbourne VIC 3051

Craftwest Centre for Contemporary Craft
Little King Street Arts Centre
357-365 Murray Street
Perth WA 6000

National Aboriginal Cultural Institute
234 Grenfell Street
Adelaide SA 5000

Craft Australia
Level 5, 414-418 Elizabeth Street
Surry Hills NSW 2010

engineering business in Melbourne, I knew when I started that I had to present myself as much as the product. In Melbourne I used to put on a suit and go to meetings to discuss problems with the different buyers because Stuart was too busy working machines. Even if you're a gardener, you don't show up to a job interview in overalls with your secateurs in your hip pocket. You still have to present yourself as a business person. You don't have to be fancy. My method of dressing is the 'KIS' method: keep it simple.

'You really have to market them yourself'

If you want to get value out of your products you really have to market them yourself. If you have a marketing manager do it, they put 30 per cent on; and then the galleries, who have to pay enormous rents, put 100 per cent on — but that 30 per cent has to come out of your pocket, not the gallery's. You can finish up slaving for nothing at all. Also, by marketing yourself, you get feedback from the gallery owner about what customers are saying about your work.

We generally go to a gallery together, and we always have a plan of attack. We've got children and friends who live in Melbourne, so we stay with them; but for people who live in country areas and don't have contacts, marketing in the city would be really difficult. You've got to put $150 a night into your costs to stay over, and you may not have even sold any of your work yet. We don't make personal contact with the galleries all the time, though. Once we've got them established, we go into Horsham once a week with a car full of packages and just send it by post.

We've just recently got work in at a gallery at Southbank, on the Yarra, but they've been struggling so we've put pieces into another couple of galleries, including one in Marysville. Everybody blames the casino — it seems to be the one thing that's made an enormous impact on marketing in Melbourne. We're very isolated here. The nearest farm is three kilometres up the road, and when we first came here, they [the neighbours] lived most of the time in Horsham. Our nearest real neighbours, then, were sixteen kilometres away.

Stuart got to know the local men through working on the farms, but it took me longer to get to know some of the women. You don't arrive in a district and find all the women showing up with fruit cake and soup. They're all busy. Most women around here work. There's a bit of a joke that all the teachers and nurses marry farmers. Apart from one that I know of, all the rest of the women are nurses and teachers.

Also, because my marketing was directed straight down to Melbourne, it took a long time for the locals to find out about the pottery. Now all our neighbours come here to buy their Christmas presents and bring their visitors to have a look. The church guild ladies will come over for an afternoon as part of their group activities. We usually sponsor one day at the local golf course each year. We supply them with all the trophies, and that means they're more aware of us.

'It's a great lifestyle'

Now the farm is beginning to build up; but for ten years the pottery has been almost our sole income. We've just done our end-of-year figures: the income we're getting off the farm covers our husbandry costs, as well as a bit of fencing and some Landcare work, but we don't get any personal income from it — the car-running costs and that sort of thing.

Because we're both home, we can get three times as much done with two people as we could with just one. It also means that at four o'clock on a sunny afternoon, if you've finished work for the day, you can go out in the truck together and do some burning off or chop wood. You couldn't do that if you had conventional jobs. So it's a great lifestyle, too.

ENTERPRISE INFORMATION

Name
Jaree Pottery

Address
RMB 5442
Horsham VIC 3401
Phone (03) 5388 1354

Type of farm
Angora goats/ Prime lamb

Year established
1986

Product
Domestic tableware/fine pottery

District
Horsham

Numbers employed
Husband and self

Market
Galleries/gift shops

Annual turnover
$30,000-$40,000

8 Margaret Currie
'Woolly cooker' manufacture

Margaret Currie is the wizard of the craft market. Her 'woolly cookers' — which can keep food either steaming hot or completely frozen for hours — always draw a crowd. She won first prize at a local 'Sheepvention' for the idea, and her business continues to grow.

I grew up on a nearby sheep-and-wheat farm, the youngest of seven children. My father had been in two wars and tended to rear us in a rather regimental fashion. Our mother had died when the three youngest, my two brothers and I, were still children. Later on in life, when I faced tough times front on, I was thankful for the upbringing my father had given me.

I married a sheep and wheat farmer. Apart from preparing the income tax, I wasn't involved much in the farm business. I always seemed too busy with the children: running to catch buses, running to sports events, and keeping the meals on the table. My interest in the farm grew once the tough times hit — the drought followed by the wet year [1982 and 1983]. Costs skyrocketed: fuel, super, and freight charges. I found myself studying land, wool, and fuel prices, which I'd never been interested in before. I'd left all that to my husband.

We needed an extra income, so I did work with the intellectually handicapped for a while and I also minded career women's children. I didn't do it for pleasure. It was necessity. Then, when my husband took sick, about ten years ago now, I wanted to spend more time at home; that's where my place was.

Around that time I came home from being interstate; there'd been a drought, and our windmill tank had collapsed. It was terrible because gardening was my hobby and my relaxation. If I was ever worried I'd just go and spend time in the garden, and I'd come in all refreshed. I missed that terribly. Then our rainwater tank collapsed, too. So that was something I had in my mind when I started the business: 'I must have a windmill tank. I must have a garden. I must have water in the toilet.'

I found myself, in my fifties, uneducated, yet fully aware that if there was to be any income I had to find a way to produce it.

'They were into this cushion cooking'

The idea came even earlier. A relative sent me these cushions from New Guinea. I didn't know what they were. She said, 'Try these out', but she didn't send any instructions with them. They weren't very attractive, so I just put them away. Then when I went to Western Australia I saw them again; they were into this 'cushion cooking' up there. A pot is wrapped inside a wool cushion which is supposed to work as an insulator, holding in heat or cold.

I couldn't believe that they would work. My son bought me one and we tried it out. Sure enough, we had this beautiful hot meal that night. But there were certain aspects about it that didn't appeal to me. It was too large and too expensive, but I was utterly fascinated by it.

I tried all sorts of different designs and so forth. I thought, if I could modify it, and put it into an affordable price range, it would be a boon for country women and busy working mothers. It saves on the fuel — our electricity bill's almost halved.

I approached the Wool Corporation, and they were very helpful and very prompt. They told me what sort of wool I would need and where to get it. It's not just any wool from the floor of the shearing shed. The wool is treated in five stages. It's treated against mould and bacterial growth. It has to breathe. It must have spring in it, and it has to be shrink-resistant.

So that was how it started. My friend Beryl Dillon helped me in the beginning, but as her working hours increased at the local hostel she didn't have time to sew or travel places.

'I can't believe how popular they are'

I handed a few of the 'woolly cookers' around to friends, and word got around. The woolly cooker is a basket with woollen cushions inside for heating and/or cooling. I thought, this is great, so then I tried wool-lined tea cosies. I can't believe how popular they are. I thought teapots went out when tea bags came in. Then I got onto woolly wine coolers, and it went from there ... pot holders, children's lunch bags, woolly carriers, or just the basic woolly cushion on its own.

I'd go to markets. I had to learn to drive with these roundabouts and lights — a bit different from driving up the dirt track. Sometimes I'd sew all night and have an hour's sleep, then I'd get up early for the market, and I might sell only one. I used to make trayfuls of melting moments, with a fancy design, so that if I didn't sell one thing I'd sell the other. That would at least pay for the fuel and my site. I'd go to the Murrabit, Donald, Swan Hill, and Hamilton markets. I felt really nervous and embarrassed when I went to my first market, but gradually word got around. The woolly cooker was the thing that got me recognition.

I looked into getting some assistance before I started the markets, but if you weren't on unemployment [benefits] there wasn't really anything offering. The most important encouragement I got was from Graham Harris, the facilitator at Donald, and Doug Sarah at Small Business Enterprise in Ballarat. They gave me confidence and ideas; the mail-order service and the parties came from Doug's suggestion.

MARKET NETWORKS

Local government and local tourist centres can advise on markets in your area. Stall holders at the markets are a good source of advice on types of markets, times, and places.

I wrote letters offering to do parties for schools, and I'd give them a basket with the cushions inside to raffle. I gave a few sets of cushions away to people to try, and they told their friends.

I didn't know how to approach the bank manager for a loan to start my business when we were in debt. I called on his wife and left a basket with her to experiment with. It must have performed well as, a couple of weeks later, when I finally found the courage to approach him, he didn't hesitate. He thought I was on a winner.

He didn't hesitate to take the first thousand dollars I earned, either. I asked him how he felt, considering it had taken me three days to earn it and it had only taken him two seconds to take it off me. With a smile, he answered, 'Got any more?' I was most grateful for the bank manager.

'My first real boost of confidence'

Then I got invited to the 'Sheepvention' at Hamilton, and that's where I got my first real boost of confidence. I took the woolly cookers and won first prize for my idea; I can't remember what section it was in. I didn't even know they had prizes. I didn't know it was such a large event. But it was a real thrill. We got a voucher for $100, and we bought a battery, of all things. So there wasn't really anything in it for me. It went into the truck.

I lacked confidence when I started. I'd always worked for other people and done as I was told. My confidence has improved. I can now get up in front of a mike [microphone] and speak, which I never thought I'd be able to. I used to have to pretend that I'd had a brandy and put on my phone voice. Although, I still prefer to speak to strangers rather than to people I know.

I meet a lot of interesting people at the markets, too. A lot of the time you meet the same stall-holders, and you form a sort of kinship. You help one another. It's great. I've met some wonderful people, and they're all out there trying to make an honest bob.

The community has been very supportive. Sometimes I'll have a second and I'll give it away, or I give them away as gifts. They always come back with glowing reports.

My husband's very good, too, and my son. My husband does the cooking and housework while I sew. I might add, I can only sew a straight line. My family can't sit down and watch television unless there's some scissors and a roll of material in front of them. They cut them all out for me. They don't mind, as long as they have peace and quiet while they work.

Some nights, if everything's going fine and the phone's not ringing, I get a really good go at it. I'm finding the older I get the less sleep I need, which is very handy. There's been times when I've sewn all night. I've had the sewing machine for 38 years now. I thought I should trade it in at one point, so I came home with a new machine. Two days later I went to the shop and got my old one back. I could go faster on it and it was stronger, although I lost money on the deal. Pretty rough, I thought.

'It's getting bigger'

The first year I ran at a loss. I went in too big, too quick. I had baskets left over, and I had to return a lot of them because I couldn't afford to pay for them. I took the wrong advice from people who didn't know enough about it. They were my mistakes, and I had to grin and bear it.

I may possibly expand in the future. I don't know. There have been people who have said, 'We'll do a brochure for you and find the market', but it'd cost me $700 and I just don't have that. Doug Sarah checked out these people for me and told me to steer clear. It's good to have somebody with the knowledge he's got to fall back on.

I've had to do my own marketing, but now I find the invitations are coming in, which has made a difference. Wherever I go, there's usually a story written up about my products. I no longer have to look for sales or travel to markets. I receive mail orders from all over Australia. Often I am in a town some distance away and I will hear a voice say, 'Hey, there's the "Woolly Cooker Lady"'. I always find that embarrassing.

When I go to the Women's Expo and farm gatherings, although there's about four or five other lines, it's the woolly cooker they're really interested in. I usually have something cooking with a really nice aroma, like braised chops with onions, and they lift it up and they can smell it. They'll come back a few hours later and find the pot's just as hot. Then, up the other end of the basket, I'll have a jellied trifle or yoghurt, or a couple of cold cans of drink. There's just a little divider cushion in the middle. It's amazing that, in the one basket, one end can be hot and cooking, and the other end, cold.

In the woolly carriers I'll usually have frozen chicken or peas. They'll be in the bags for hours and they just do not thaw. My most common comment is, usually as the customer scratches their head, 'That bloody thing works'. They can't believe it until they try it — that a meal can cook in the boot of a car as you travel.

The most unusual way my cookers have been used, to date, is to rear baby ducklings that were left motherless. The wonder of wool still amazes me.

ENTERPRISE INFORMATION

Name of enterprise
Colonial Nu Wool Products

Address
RMB 448
Birchip VIC 3483
Phone (03) 5492 2671 or (03) 5493 7612

Type of farm
Grain/sheep

Year established
1993

Product
Woolly cookers

District
Wycheproof

Numbers employed
Self and two part-time staff

Market
Australia-wide

Annual turnover
Unspecified

9 Cicely Fenton
Conservation farming

Thirty-eight years ago, Cicely Fenton and her husband started planting trees to create shade on their hot, dry property. The locals thought they were mad. Now the 'agroforest' attracts a constant stream of both Australian and international visitors interested in land care. The family received the Australian Forest Growers' state award in 1993 and the national award in 1994, and Greening Australia's award for outstanding achievement in 1995.

John and I will have been married 38 years tomorrow, and we've been on the farm all of that time. When we first came here, it was a square house on a square hill — hot and dry. Both of us had come from a background of trees: John from Hamilton and me from the redgum country at Vasey. We felt it was a necessity to have some shade and shelter; planting trees has been a continuing thing since then.

The neighbours, over the years, have considered us quite odd, because we wanted to plant trees and make the environment a more pleasant place to live. We got no community support. We were 'way out', 'most peculiar' ... It took 25 years for people to realise that planting trees is worthwhile.

'They said pine trees wouldn't grow here'

When we started doing agroforestry we got absolutely no support from any government departments. They said pine trees wouldn't grow here.

To support what we were doing, John did some planning work with Tract Consultants, a large landscaping firm in Melbourne. Then he formed his own company from home, Rural Trees Australia. While all this was taking place, I was running the farm.

We had a chap working for us from 1958 to 1966, but after the 1966 drought he had to go, and the kids had to bog in and do it. The eldest one was only three or four. We couldn't afford to have anyone working for us, except for the contract work of shearing and fencing.

For some strange reason, the children all still want to be farmers. My daughter's now married to a farmer, and the twin sons are both married to very capable girls and are farming. The youngest son was farming until he was hit by lower prices and high interest rates. That finished that.

Four or five years ago, a friend who owns a big commercial tree nursery at Mt Dandenong asked if I would I go down and look after the house, and work in the nursery and keep an eye on things, while they were in America for six months. The neighbours thought we had split up! Off-farm income does help, as on-farm diversification is not always easy.

'It has world standing'

Now 'Lanark' is considered a showplace of the whole district. Because of what we've done, we have groups visiting from universities and agricultural colleges all over Australia. We've had various overseas universities here as well. I suppose you could say it has world standing. John gets paid by the hour for talking to visitors, and if I feed them I get paid for that, too. So it adds to the farm income.

We were doing it [receiving visitors] *gratis* for a long time. We've got a long-standing relationship with RMIT and their landscape students. Then in 1993 the Professor of Landscape, Jim Sinatra, said to us that we had to start charging people. That's how it eventuated as a financial venture. We thought we were just doing everyone a favour by helping them get the land care right.

The only government support we got was from Rod Bird with the

Department of Agriculture in Hamilton. They were desperately looking for areas where they could do some research, and I suggested to John that we give them an area here. So now they plant it, they maintain it and, when their research is finished, the trees become ours.

We also have a unique situation here: the Department of Natural Resources has released Victoria's most endangered mammal, the Eastern Barred Bandicoot, for reintroduction into the wild. This was the first release on private property, because areas we closed off to grazing 30 years ago are now ideal habitat for this special little animal. They pay me to assist with the monitoring and to do fox-baiting. Fifty have been released over four years and last year a count revealed ten 'clean skins' — bandicoots born and bred here on the farm.

This is all part of our venture into education. People from all walks of life are now keen to care for the environment, and to learn how to do things related to land care.

'We're looking at a range of enterprises'

We now work directly with the New Zealand Forestry Institute at Rotorua, and we try to plant at least one hectare of pines every year. By the time we've planted up to 24 hectares, the first of the trees that we planted will be milled. This should bring in extra income for the farm.

The income that we've made from diversification so far is probably only small. But now, with David [the youngest son] running the farm, John and I are able to put our time into talking to interested groups who visit. Really, the 2,000 acres here is only a living area. It did support us and four children, but now it's going to have to support us and David's family as well.

I'm starting to use the reject trees that we've planted to make walking sticks, but I haven't started selling them yet. I also have a small herd of stud Shetland ponies which I sell, and a patch of flowers that my daughter-in-law and I have worked very hard on. Unfortunately the last crop failed due to too much rain. Really, we're looking at a range of enterprises to try to increase our income.

One of John's pet hobbies is trying to get the councils to de-rate the waterbodies that have been fenced off and re-vegetated for land care and trees. At the moment, they're not interested. The local councils are all broke and they're after every penny they can get.

The long-term plan is to survive, one way or the other. It is beautiful here, but it has been a big cost over the years. Really, the amount of income we get from showing people around is minimal. But it's a lovely place to live. Three years ago we planted 600 oaks for our great-grandchildren. Nobody else has done anything like that.

ENTERPRISE INFORMATION

Name
Lanark

Address
'Lanark'
Baranxholme VIC 3302
Phone (03) 5578 6243

Type of farm
Sheep

Year established
1958

Product
Farm forestry and educational tours

District
Southern Grampians

Numbers employed
Family

Market
Tertiary institutions, landcare groups, and primary schools

Annual turnover
Under $5,000

10 Anne Gready
Book-keeping

When Anne Gready was told she wasn't much use on the farm, she decided to find out what she was good at. It turned out to be numbers. After four years of study, she and her partner Gwenda Grigg now operate a successful book-keeping and secretarial service for farmers and small businesses throughout the Wimmera and Mallee.

I grew up in Horsham, but I may as well have lived in the centre of Melbourne because I knew nothing about farming. I worked in a bank after I left school, and then I met my husband [Philip] and moved to Lubeck. We live in the township, but the farm's only a mile out of town. There are five houses in Lubeck, and ours is the one on the right!

'I knew I wasn't going to be a housewife'

From the day we moved here, I knew I wasn't going to stay home and be a housewife. It's a big house and I hate housework. I remember the evening we came home from our honeymoon, and Philip walked in the back door — in the house he'd been in all his life — and said, 'What's for tea?' We had frozen peas for our first meal because that's all I could cook. The chooks ate well for a long time after that while I tried to master these things called cakes. My specialty now is pavlova and ginger sponges; through pure determination I beat the dreadful things.

So I went out to work on the farm with Philip. My father-in-law thought that I was too skinny to be any good on the farm, that I'd only be good for holding something or getting a cup of tea. It was true that I couldn't physically cope with a lot of the work ... and I didn't like the heat ... and I get hayfever ... and I was allergic to dust ... and I couldn't drive.

To top it all off, when I was expecting my second baby in 1978 I had a crushing experience at a clearing sale. An older man from the district, who was sitting down next to me having a cup of tea, said, 'You belong to Philip, don't you?' I said, 'Yes, I belong to Philip'. And he said, 'You were a nurse, weren't you?' I said, 'No, I wasn't'. And he said, 'Oh, were you a school teacher?' I said, 'Well, no, I wasn't actually. I worked in a bank'. And he said, 'Oh, and I thought you were something useful'. With that, he got up and walked away.

When I got home I shed a few tears over it, being a bit emotional at the time. But I thought, no, I'm not useless. I've got to be good at something. I couldn't cook or do much else at the time, so I was driven by the desire to find out what I was good at.

At that time I still wasn't in the paid workforce. Then things got particularly tough during the drought [1982-1984]. We'd just had our third child, and I was told I couldn't pick up my baby any more because I had fibrositis. He was a lug of a baby, too. So I was in a bad way physically, getting very frustrated with life and not fitting in with the community at all well. Still trying to master the sponge cakes and everything else.

But I realised that when your back's against the wall and you can see no way out, there's only one way — and that's straight ahead. My husband felt the same. So we dug our toes in, and I decided to forget about everybody else and concentrate on us.

'I wasn't a freak after all'

It was after the fourth child, Kate, was born in 1985 that I decided that I had to do the farm books better. I was budgeting, but I felt there had to be a better way of doing things, and that that's where I could contribute. I heard about a six-week book-keeping course at the Wimmera TAFE in Warracknabeal, so I enrolled. With one baby under the arm and the other three children at primary school, I used to take off every Tuesday to do the course. It turned out not to be six weeks but four-and-a-half years, and at the end of it I had my Certificate of Business Studies (Agricultural Secretary).

It was difficult because Warracknabeal is 77 kilometres away, and I had to put Kate into creche. I'd never put any of my children through kindergarten even, so I found that very difficult. My in-laws thought the course was a waste of time and money, and quite silly. My husband always found that Tuesday just happened to be the day he needed a hand to drench sheep or something. So for the first twelve months, every Tuesday was a terrible inconvenience. But I went anyway.

And I loved it. I loved mixing with other women and finding that there were other women like me who hated housework, had a strong desire to be involved in the farm, and loved book-keeping. I wasn't a freak after all. It was a wonderful feeling. I felt this was where I belonged, and I excelled at it. I'd hated maths at school, but I was just breezing through this course.

I actually ended up doing an Associate Diploma in Accounting at the same time, so I was getting very involved in that side of the farm as well.

'I'd found out what I was good at'

I was very proud the day I was able to say to my father-in-law, 'See this money that's been paid into the bank — that's the refund on the overcharge in interest that I picked up on. I earned that money.' I was really pleased that I had finally contributed something to the farm.

I'd found out what I was good at: book-work and talking to banks. It took off from there. After I'd got a couple of years of the course under my belt, I could see there was a need for better book-keeping in the community. I could also see that there was money to be made out of it.

Around that time, I found out that there was a woman, Gwenda, who was doing the same course in Rainbow and who had similar ideas to mine about making a business out of it. When we graduated we met up, and within a couple of months we'd started up the Farm Secretary Service.

I had $3,500 to get us going. I thought I was loaded; but it wasn't anything near enough, so we advertised on a shoestring. We got as much free advice as we could about how to advertise really effectively rather than just throwing all your money at it.

We went to David Furniss who was the REV facilitator in Warracknabeal. He was a big help to us. We also went and saw Sergio Guerra, a solicitor over in Donald, and he gave us some legal advice. We used Rob Taberner, the rural counsellor. We used as many people as we could.

But advertising was still expensive, so it took us a long time to get accepted. It doesn't happen any more, but men used to walk past us at field days and laugh. They'd say, 'There's those silly secretaries'. But every year we'd be at the same site and then they'd say, 'Oh, they're still there. They must still be in business'.

'They think it's great'

Now we've got to the stage where farmers say, 'You're like shearers. We pay you, you do the job, and we're happy at the end, so we get you back again'. They think it's great.

I also teach the book-keeping units of the Office Administration courses at the Horsham Learning Centre, and I do the financial modules for the Small Business Management courses they occasionally run. At one of these sessions recently, a businessman asked me, 'What happens when you can't tighten the belt any more?' I said, 'Sell the belt and go without.' I've become very money-oriented. It's a terrible thing, but with every downturn you get into survival mode, and you're driven by that.

The community understands now that I am different, and they are starting to accept me. If there was a farmers' meeting out in the paddock somewhere, I'd be the one woman going along. Now other women are coming along, too. I was also one of the first women to get really involved in the local fire brigade, so I was also accepted by the local men.

'It's taken pressure off the farm'

The income has been paying a few farm bills, and it's picked up a few of the expenses. I've also used the money to buy a brand new car and pay for its running expenses, as well as the business expenses and all the clothes and shoes I could desire. It also helps out with the kids. It's taken pressure off the farm.

We've got a large enough clientele now. I really don't need too many more. I have a natural drop-off of clients because I'm always training farmers to do their own books. It's lovely to see them go off on their own. Then I pick up more. It's going along quite nicely. It will probably increase, but I like to have a balance. I think it's essential for our business that Gwenda and I are still involved in our own farms because it means we understand our clients' needs. Anyway, the accountant keeps saying he doesn't want it to get too big because of the tax side of it.

It's been a long haul. There's an old saying, 'Don't give up your day job'. Well, I didn't have a day job to give up, but it hasn't been an easy juggle. It's a matter of working out what's important to you, what's important to the farm, and what's important to the family. I really do think I've found a balance between the three.

That old man from the clearing sale is six feet under now, but I'm still here and I'm doing something useful. My husband says he doesn't think we'd still be farming if it wasn't for what I've done with the finances. I'm inclined to agree with him. I think I have done a lot to keep us going. I've always tried to run the farm as a business. I'm not as emotionally attached to the land as my husband is. I've always tried to say, 'Yes, I love the farm, but it's a business'. That was the tack I took fifteen years ago, and I think it's paid off.

'We have a long-term vision'

I had the impossible dream: I wanted to work; I wanted to have my independence; I wanted to contribute money to the farm; I wanted to have my money and spend it, too; and I wanted to be here for the children after school. I also wanted to be on the farm at the important times like shearing. I had to be my own boss. So I needed a magical job that would pay lots but let me have days off whenever I felt like it. And I found it. This job lets me do all these things. It's turned out to be the perfect job.

We have a long-term vision, Philip and I. It involves a white four-wheel drive ute with a little aluminium boat on the top and a campervan hooked on behind. We've got a road map heading north, and we're not coming back. That's our vision. But there are a lot of things that have to happen first. The four children have to leave home and become established in whatever they're doing, and there's got to be somebody working the farm. My business is a real part of achieving our vision.

ENTERPRISE INFORMATION

Name
Farm Secretary Service

Address
RMB 6209
Stawell VIC 3380
Phone (03) 5359 6219

Type of farm
Grain

Year established
1992

Service
Book-keeping and secretarial service

District
Wimmera

Numbers employed
Self in Wimmera and partner in the Mallee

Market
Farmers and small business

Annual turnover
Over $10,000

11 Gwenda Grigg
Book-keeping

Gwenda Grigg started a book-keeping course because she wanted to improve farm finances. It grew into much more than that. With her business partner, Anne Gready [see pp 76-83], Gwenda now operates a farm secretarial service that is expanding at a rate of 25 per cent a year. She intends to use the extra income for lifestyle improvements.

I grew up in Rainbow; my father was a general store keeper. I've lived here all my life, except for my final year of school when I went down to Hamilton. We didn't have matriculation, as it was called then, in Rainbow. I had a wonderful time in Hamilton, and wasted my educational opportunities because I enjoyed myself too much.

I went into the bank at Rainbow after I finished school. I really enjoyed that; it was good fun. I probably could have gone to work in Melbourne, but I guess I was just more comfortable here, and the opportunity never really came up.

I married in 1978, when I was 21, to a Rainbow lad who was a farmer and who had grown up on a farm. Geoff had gone away for a year to try out Melbourne, but he came back and chose to make farming his career. When we married I stopped work immediately. We had two children in the next five years.

'I decided to buy myself a camera'

In about 1983 I won $100 on the 'scratchie' tickets and decided to buy myself a camera. I was going to buy one of those ones where you take the photo and it prints immediately. But Geoff said, 'Don't be silly; it's a waste of money'. So with my $100 I bought a $300 camera — it was my birthday about the same time.

I really enjoyed photography, so a few months after that I decided to learn a bit about black-and-white photography and how to develop [photographs] myself. Over time, I bought a better lens for my camera and worked out how to photograph reasonably well. We had a spare house on the farm, so I converted the kitchen area of that into a darkroom. I put black plastic up in every window.

I started reproducing photos of local people and places in order to compile an oral and visual history of Rainbow. I developed this hobby for my own interest, so I was copying family photos and also going around town interviewing elderly people and borrowing their photos. It was really good fun and I learned a lot about the history of Rainbow.

Over the next eight years I built up a collection of about 500 photos, which is quite a valuable resource. At the moment they all just sit in files at home, and anyone who hears about them can come out and have a look. Some of my reproduction photos have been used in the Shire of Dimboola and also the Shire of Karkarook history books.

I financed that hobby by letting people know, who were interested in family history, that I could reproduce their photos for them. I got quite a lot of business — mainly around the Rainbow area, but also from people further away who would send photos to me. The income from the photos didn't pay for holidays or anything, but at least it kept my hobby going. For the hobby to be able to finance itself has been very important to me. Then anything that I wanted to buy, mainly more camera equipment, that I thought might have been a bit extravagant, I would use that money for.

'Rainbow started a learning group'

About ten years ago, when I was in my early forties, Rainbow started a learning group. They ran a business course for farmers, which was just two or six weeks. The group decided from that, in conjunction with Wimmera TAFE in Horsham, that there was a real need for a much broader-based farming education — book-keeping and that sort of thing.

So TAFE developed the Certificate of Business Studies (Agricultural Secretary). They came up to Rainbow once a week for four years, and at the same time they were going to Warracknabeal, which is where Anne did her studies. They supported the course with child care. That was extremely important, because without that we wouldn't have got the women. Only two men took part in the course.

We needed ten students to get the course started but, for various reasons, people were dropping out. We very quickly realised that we couldn't sustain it for four years with three or four people. So when we were sitting around watching football, those of us that were doing the course started talking very enthusiastically about it to other women at the football. We ended up with about another 32 people, and they did parts of the course. Some just did the communication skills, others did computer subjects, and some did the accounting subjects.

I loved studying, once I got into it. It was a real challenge, but I just loved it. The sense of achievement and self-esteem I got from my success — because I did have success — was very important to me. I've really built up so much more self-confidence in dealing with people. I got involved with our local learning group at the same time as starting the course, and I've been on their committee ever since.

'Just forget the title!'

I'd like to really stress to people that are thinking of taking on study, but who are put off by the title of the course they're interested in, just forget the title! Forget about the fact that it might take two years, or four years, or six years part-time. Every course starts at the beginning, so everyone can achieve something beyond that level. Even if they find it too much, they've still gained those first few subjects they were interested in; or they might find that they really love it and go on with it. That's what I gained from it, and that's what I'd really like to encourage other people to do.

In the end, three of us graduated at Rainbow, but the other people all gained something from it, and some then went on to do other courses in Horsham. It was very valuable for all of us; we all gained self-confidence from it.

The Rainbow students knew this other group was meeting over at Warracknabeal, and we decided that we should all get together for our final session and just get to know each other. One of our tutors, Nicki Berrisford, knew that I wanted to go on and do something a bit more than just our farm. I started out just gaining skills to improve the book-keeping on the farm, but it became much more than that. Nicki also knew that Anne, over in Warracknabeal, was thinking of doing something more with it. So she said, 'You two should get together and talk'.

'Within three weeks we had decided on a business name'

We did. We talked during that final session of the course and agreed to meet a week later and discuss it further. Within three weeks we had decided on a business name — Farm Secretary Service, because we had always just called it the farm secretary course. The Wimmera field days were coming up very shortly, so Wimmera TAFE offered us space in their on-site van to start our business. They were very proud of us; the fact that we had finished this course and that we were starting up a business. That got us up and going: we had three months from when we first met to our first day of business.

Along with the support from TAFE, the *Wimmera Mail Times* gave us editorial copy and took photos of us. We put an ad in the field day supplement, and that got the word out that this new business was on the go. We had some inquiries over the next six months, as a result of the field days; but when people come and talk to you, they often don't follow that up with a phone call for two months, or six months, or even two years later.

I did other work while the business was getting off the ground. Not the last Census, but the one before that, I got a job as a Census collector. Then, after that, the Bureau of Statistics needed an interviewer in our area. I decided to apply for a job, and I was fortunate enough to get it. So I worked for the Bureau of Statistics for five years, and travelled quite extensively over the Wimmera and Mallee in that time.

It wasn't full-time work, but I was on the road a lot, and that was often quite tiring. When I started the job, it was only three days a week, once a month, but that built up. Once you become an interviewer you become more skilled and you get offered a lot more surveys, so by the end of the five-year appointment I was working more than I wanted to. So when the job with the Bureau finished, I was really quite happy. I really enjoyed it, but I had had enough, and my own business had built up quite well by then.

I'm happy with the way the business is going. We've still got room for more clients, but sometimes you're working flat out for three weeks on end, and then you have a few days at home and it's a real joy.

Our clients invite us to their farms. They've either read about us in the paper, from our advertising, or one of our current clients has been talking to their neighbours. They ring up and ask if we can help them and we say, 'Yes, we can, in some way'. We don't offer a specific service. We are just as happy to teach clients how to do their own books as to go in and do all the book-keeping for them.

'We do a lot of one-off sessions'

In February, March and April, when grain farmers are budgeting for the year ahead, we do a lot of one-off sessions. Hopefully they've had a good harvest, had a bit of a holiday, and they're now ready to plan next year's finances. We'll spend a morning doing a budget with our client, and that's all we do for them.

Budgeting is just a tool. No-one sticks to a budget — unless the bank is really, really strict — but they are a good guideline. It is also important to get your books up-to-date, at least every three months if you can, so that if there's going to be a problem you can see it one or two months ahead. Then you can do something about it: go and see your banker and arrange to have finance for it, or look at forward-selling something that you hadn't planned to sell till later on. Many farmers now have computers for their children at school, so if they're going to do book-keeping and they have got a computer, it's best to start them off on it. There are simple programs available that will meet their basic book-keeping needs, and that, in turn, will meet their accountants' needs. If you want to get into budgeting yourself, and really good financial planning, you need a better quality program. But I always suggest that people start off with the basic accounting program and then, when they get frustrated with the inadequacies of that program, at least they've got a better understanding of what they need and what they want.

At Last: Collecting the Statistics

In 1995 the Australian Bureau of Agricultural and Resource Economics (ABARE) published the first major set of statistics on the role of Australian women on farms. The data was collected as part of ABARE's surveys of family broad-acre and dairy farms, and therefore did not cover the full spectrum of Australian farming. It nevertheless gave an insight into the demographic characteristics of the women as farmers and the key social issues concerning them.

The data showed the diversity across industries of the working patterns of women on farms, and the various perceptions of the major issues affecting the women and their communities. Results differed according to the women's age, the industry they were in, and the broad-acre zone of the farm business.

Some women were involved in the farm business as sole operators and decision-makers; others shared these roles and also undertook household and family duties. Some worked on the farm in peak times; others did home duties and were not involved in day-to-day farm operations. Others had full-time off-farm work.

Although the survey information established for the first time a database to assist the Australian government in the development of future policy, the information raised more questions than it answered. Future studies were planned, broadening the methodology and sampling, to give a more complete picture of the increasingly valued role of women in Australian agriculture.

For more information about this survey, see ABARE Project 1088. *Australian Bureau of Research Economics. Canberra 1995*

We also do some work for small businesses. Small-business people need the same sort of services and we can do it for them, but 99 per cent of our clientele are farmers.

'I want to keep my business growing'

The business is building up by about 25 per cent a year, which is good progress. It's a pace that I can handle. I'm always thrilled when a new client rings up out of the blue and I've got a new job to do. Sometimes it leads to just a one-off job, and other times it leads to seeing that client either monthly or every three months. The clients that are local, that can either drop their books in to me or I can visit on their farms, tend to be monthly; but not all of them.

I'm still enjoying it so much that I want to keep my business growing. Perhaps if I get to a point where I can't take on any more clients, I'll concentrate more on tutoring, getting people to do their own books.

I've actually got two sorts of clients. I've got one type that have never done books, but now see the need for better financial management. They need their books done and they find that getting me to do it is very cost-effective. The other type of clients are people who have always done their own books, but now, through changed circumstances, they haven't got as much time to spend in the office. By getting me to do the basic book-keeping, they're more free to analyse the reports and work out where they're going.

Anne and I didn't get any assistance from local government, but we didn't ask for it. Anne and I just set the business up. Because we're both on farms, we already had an income, so we weren't reliant on this income for everyday needs — although it has been handy to have extra income. Mine probably goes more towards lifestyle things, like holidays or improvements to the house.

But then, I guess with any farm woman, we have the constant problem that when I build up a bit of money in a bank account, there's an overdraft in another account. It's very hard, then, not to put money over into the farm account. After all, it's the best interest rate going. Of course, once it goes, it never seems to come back the other way; but I'm saying that one day I'm going to get that back and make improvements to the house.

We're now planning major alterations to our house. I'm now 51, and we've reached an age where we've got to make decisions about whether we stay or whether we move out. Neither of our children are on the farm and are not likely to come home. Geoff's decided he'd like to stay on the farm. My business has built up in my area, so I'm happy with that.

Taking up adult education has been a major influence in my life, and there are many opportunities for isolated women to start a course these days. It's much easier to study if you can take part in a class, rather than on your own at home, but either way I say, 'Go for it'. You can only benefit from wherever it may lead you.

ENTERPRISE INFORMATION

Name
Farm Secretary Service

Address
PO Box 181
Rainbow VIC 3424
Phone (03) 5395 1171

Type of farm
Grain and some sheep

Year established
1992

Product
Book-keeping and secretarial service

District
The Mallee

Numbers employed
Self in Mallee and partner in Wimmera

Market
Farmers and small business

Annual turnover
Under $12,000

12 Heather Hamilton
Upholstery & emu farming

Heather Hamilton was a 'townie'. When she married and moved to a wheat farm, she had to learn fast. To earn extra money to buy the family clothes, she taught herself how to upholster tractor seats. Then she taught herself how to raise emus. In addition to all this, Heather continues to fight to keep the district's pre-school open.

I was a 'townie' before I got married and I had to learn a lot fairly quickly — the hard way. My father bought a 3,600-acre farm and asked Ross [Heather's husband] to run it because he comes from a cereal farm. This year [1997] we started leasing 400 acres of the property from my father.

'I do a lot of tractor and ute seats'

I got into upholstery when my first car needed the back seat re-covered. I was going to get it fixed and my father said, 'Do it yourself. Don't be so lazy'. So I did, and I've been doing it ever since.

I do a lot of tractor and ute seats for farmers. One day a carrier brought over a chair and dumped it off and said, 'This is for Haeuslers Farm Machinery of Sea Lake, they want it fixed'. I've been doing work for them ever since. A lot of the farm women want their kitchen chairs re-covered. I've even had some townies come out, wanting me to do their work. I don't have to advertise; it's just word of mouth. They find out I'm here.

When I'm having a good year, we probably get an extra $5000 or more. It pays for our clothes and luxuries. I like to go to the Murrabit market, which we can't always afford. With Luke [Heather's son] —

well, you're always wanting things with kids, aren't you? And it's amazing the clothes Ross goes through.

'Emus are as iffy as they come'

When I got pregnant with my second child, I stopped doing the upholstery because tractor seats are so heavy. The emus were my father's idea. He said, 'You need something extra in case it's a bad year'. But the emus are as iffy as they come; we don't know if they're going to be any good.

When we first bought the chicks, to raise them for meat, we had to lock them up at night and only feed them a little in the morning and a little at night. That was my job. I had to count each chick because you could only have so many in a pen or else they'd suffocate themselves during the night. You've got to turn on the hot lights and, if it's going to rain, you have to put them back in the shed. I did all that. We decided the chicks weren't worth it after all that, not when you pay $80 a bird to get them killed and don't know when you're going to get paid for them.

So then I advertised in a local newspaper to sell the emu eggs. I go and collect the eggs — which a lot of farm women around here won't do because the emus chase them. At first I would walk around collecting them, until they started laying too many for me to carry back. One day I was walking back and my pants were nearly falling down because I had all the emu eggs in my pockets. So Ross fixed up a

Day Care for Farm Children

Farmers, with their irregular work patterns, and living outside populated areas, have seldom found child care accessible. However, in 1997 the Australian government took up the principle of flexible payments for low-income families so they could choose their own child-care provider. The government also recognised the need for additional funds for rural services. Child care is now 'attached' to the child, not the centre. This new flexibility may suit farming parents. For example, they can now pay a neighbour for long hours of child care during the few weeks of the annual harvest.

small four-wheeled motorbike for me and it pulls a big plastic bin behind it.

You soon learn that if an emu has his head down low and he's coming for you, he's going to have a go. They only put their heads up high when they get close enough to kick you. I just go around with a plastic rake, that's all I need, and hold that up higher than them. They think I'm bigger, so they leave me alone.

We've got a rack in the carport that keeps the eggs fertile if someone wants them for incubating purposes. Otherwise, I blow the eggs for a woman in Sydney. She makes music boxes and carvings and all that. To blow them, I drill a little hole at the top and one at the bottom and then hit them with the air compressor. I started off by blowing them with my mouth. I was getting dizzy and my head was spinning. Then Ross comes home and says, 'Why don't you use the air compressor?' And I say, 'Why didn't you come home after the first one?' After I've blown them and cleaned them up, I roll them in foam and tie them up like bon-bons.

The only thing Ross does with the emus is mill their feed. The rest I do. I didn't get advice from anyone. I haven't even heard of places that help you with those things. I just did it.

It's good because I'm still free to help Ross around the farm and take care of Luke. I mainly used to do the upholstery when Luke was having his afternoon nap. I'd have the baby monitor in the carport and I'd be busy sewing away. When he got older I was able to do a little bit, but a three-year-old wants a bit of attention, you know? Which is fair enough — especially considering we've no neighbours for him to play with. There's just him and me all day long. Sometimes I collect the eggs while he's still having his nap, but he loves coming with me. He sits in the front of the motorbike and I get him to look out for the eggs, and he's so happy.

'It's a struggle to keep the pre-school open'

I'll have the baby at the end of September and next year Luke will go to pre-school for two days a week. But it's a struggle to keep the pre-school open. There are fourteen kids there at the moment, and we've got to raise $7,000 each year just to keep it open.

When we have a street stall, you're baking for a full two days beforehand. And we have a goods-and-services auction. We go round to all the businesses in Swan Hill and ask them to donate something. Some farmers go out and cut a load of fire wood — anything. Last year I did a lot of free upholstery as a fund-raiser. I've been helping the pre-school since Luke was one, to keep it afloat. It's not easy.

This year we've had a big kerfuffle; they've [the Education Department] been trying to close it down. They wanted to re-class us, and it was going to cost $25,000 to do all the updating they wanted. We wrote back to them and said it wasn't possible. We said there's three venues we can use: we can stay in the hall; we can move to the primary school into a portable — but their toilets are no good, they've been trying to get funding for six years; or we could go to the mud hut at the football grounds. We asked them to send someone down to advise us. They haven't turned up yet; we're still waiting.

'Next year's going to be crunch year'

There's a block of land coming up for sale nearby and we're debating whether or not we can afford to buy it. It doesn't look like a good year this year. But we've got sheep as well as cereal, so that helps. If the crops fail, wool's on the way up. And Ross is into yard-dog trials. That might be something else, you never know.

I plan to do the upholstery again next year whether the emus are a worker or not. It would be hard to expand the upholstery business and employ someone else because we live such a long way out. We're in the middle of nowhere.

At the moment the emu-egg sales are just paying for the emu feed. Next year's going to be crunch year. If the emus aren't selling, we'll just have to get rid of them and find something else. You have to — the government's got their hand out so often. But there's always something comes up, and farm women know how to grab hold of it quicker than anyone.

ENTERPRISE INFORMATION

Name
No registered business name

Address
'Fairview Farm'
RSD Waichie Road
Swan Hill VIC 3585
Phone (03) 5039 6346

Type of farm
Grain and sheep

Year established
1994 [upholstery]; 1995 [emus]

Product
Upholstery/ Emu eggs

District
Ultima/ Swan Hill

Numbers employed
Self

Market
Local farmers & machinery sales outlet [upholstery]
Craft shops & rural newspapers [emu eggs]

Annual turnover
Under $10,000

ICE-CREAMERY

13 Margaret Hitchings
Biodynamic ice-cream making

Margaret Hitchings makes biodynamic ice-cream in the bluestone buildings of her nineteenth-century farmhouse near Colac. Profit from the ice-cream has kept the family's dairy farm afloat through several dry, difficult years. Now, with growing demand from the Japanese market, the business is ready to expand.

I was born on a dairy farm in Rutherglen. When I was five, Dad purchased a farm down here near Colac and set up a sheep and cropping farm.

I became a primary school teacher and then met my husband — a dairy farmer. I kept teaching after we got married, but I also helped out with the milking after school and with most other jobs during the harvest.

'We began producing biodynamic milk'

In the late 1970s we decided to have the farm certified biodynamic. It was a pretty radical thing to do in those days. We were interested in nutrition when the children were little, and we thought, well, how can we farm without using any chemicals? So Mum and Dad came across a biodynamic farm at Shepparton, and we went and had a look at that and learned about the ins and outs of it all. We went to a few biodynamic conferences and started up in 1980.

No-one else was doing organic farming in the Colac area at that time. We began producing biodynamic milk, but there was no market for it. We were just selling it to Bonlac like any other milk. It was mixed in with ordinary milk for products like cheese.

Because of the organic methods, production went down at first. There was a bit of a lull; it was depressing at times. Many a day my husband would say, 'Oh I'd love to get that super spreader out and go crazy with it'. Of course, I'd say, 'Give it a little bit longer'. Fortunately we persevered. I had to keep teaching and I suppose that was the only thing that kept us going, having that little bit of extra income. So we hung in there, all the while thinking about how we could get our biodynamic milk out to the public to get the full benefit from it.

I thought I'd ring the VDA [Victorian Dairy Authority] to find out about the chances of bottling and selling it, or putting it in the long-life packs. I got an answering machine and gave them all the particulars, and I didn't get a reply.

'They said, "Why don't you do it yourselves?" '

I more or less stumbled across the idea of ice-cream, through a friend of mine who was making it — conventional ice-cream. We thought of supplying this lady with our milk so that she could make biodynamic ice-cream. We went and talked about it with her and then we talked with our friends and they said, 'Why don't you do it yourselves?'

Because I was teaching full-time, that left Gary to go and do the ice-cream-making course at Gilbert Chandler down in Werribee. We were lucky enough to get some money through the teachers' union to set up the factory. It took a considerable amount of money to get pasteurisers and batch freezers, and we had to completely re-line the old original laundry. We had to build a freezer room and get a special container made up for the back of the Land Cruiser to carry the ice-cream to market. So it was quite costly.

We started the ice-cream in 1990. Then, at the end of 1993, I had to forgo full-time teaching and return to part-time.

I tee up all the markets and do all the book work, and sit here every night laboriously putting on labels. That's a family concern at night, when we're watching the news.

Going to shows and expos gets you off the farm for a bit. We laugh sometimes and think, well, that's our holiday, our recreation. It pays for the holiday, what you sell.

'I don't want to just milk cows for the rest of my life'

When we were first setting up, we went to the Department of Ag[riculture] in Colac and they were very helpful with information as to what all the regulations were that had to be adhered to. But as for financial help, we had to do that ourselves. We relied on the money from the teachers' union.

I'd say it was probably financial need, more than anything, that spurred us on to start the ice-cream business, because the late 1980s and early 1990s were a little bit lean with dairy. The other reason was probably just the challenge. I can remember Gary saying, 'I don't want to just milk cows for the rest of my life'. Here we were, we had a wonderful challenge ahead, and we thought, we're not going to let it beat us.

Fortunately the reaction of the community was quite good. They were all keen to see whether we hung in there. It's funny now — six, seven years later — a few friends are saying, 'You're *still* making the ice-cream?', as if to say, well, we didn't think you'd last that long. Probably because of all the publicity, they almost respect what we've done now.

And the ice-cream has helped pull the farm through. Hugely. Even this year — we've had such a dry year — we often say to each other, if it wasn't for the ice-cream we'd be really struggling. The ice-cream seems to be paying for quite a lot of the farm bills.

ORGANIC GARDENING AND ORGANIC PRODUCTS

There are several organisations which set guidelines for growing and producing products. You can find a large list of these in the 'Further Information' section at the back of this book. Another valuable source of information is your local department of agriculture.

'At a crossroads'

We're at a crossroads now where we're ready to expand but our machinery's not quick enough. We produce 150 litres a day and it can *just* keep up with the current demand. Sometimes it's a struggle. But now we've got three companies in Japan wanting it; they're talking a tonne a day. Well, there's no way we could possibly keep up with that with the machinery we've got.

We also need to look at improving our packaging, so that the labels are already on it. We did investigate that not long ago, but the big companies that produce that kind of packaging only do it in huge amounts.

We'd also like to move the business closer to the road so that tourists can come and have a look without coming right into our home. I work in with other local farm women, like Wendy Dennis who makes woollen products, so that if a bus goes to her place it will also come here and vice versa.

If we're game to go ahead with the expansion, we're going to have to get lots of help with the debt. So that's challenge number two. I probably won't be able to fit in any teaching; marketing will be my main job, but I enjoy it. You get lots of contacts and meet some really nice people.

I'm looking forward to it in a way, but it's a little bit scary. But I guess if you don't have a go, you'll never know.

ENTERPRISE INFORMATION

Name
Irrewarra Dairy Products

Address
RMB 1085
Irrewarra VIC 3249
Phone (03) 5233 6335

Type of farm
Dairy farm

Year established
1990

Product
Homestead biodynamic ice-cream

District
Colac

Numbers employed
Self, husband and three children

Market
Australia and overseas

Annual turnover
Approx. $70,000-$80,000

14 Krys Howard
Gift hamper production

It was a 1990s' version of the 'magic pudding'. When Krys Howard boiled up a few plum puddings for the school stall, she had no idea she'd soon be filling an order for more than 1,500. Krys and her husband Neil now specialise in gift hampers featuring a range of local products. The business has helped them deal with the inevitable down-cycles in dairy farming.

We bought the dairy farm in 1983. The year after, Neil left his job with National Parks to go full-time on the farm while I kept my nursing job three days a week.

In 1986 we were greeted with the news of a baby, which was pretty exciting because we'd been told we wouldn't have any. While I was in hospital having Dale, Neil took a fall off a horse and fractured his spine. I came home with the new baby to an injured husband, and my workload then changed dramatically.

Over the next four years, Neil's health went up and down. Finally, when the second baby came along — we were told we'd *never* have another one — they [doctors] decided that they would be able to operate on Neil. He went to Melbourne and had some surgery which put him completely out of action for twelve months. During that period, we had a lot of help on the farm from friends, and I nursed on and off.

'From a little project to a big project'

The puddings started the year Dale started school, which would have been 1991. I was asked to do the usual mum stuff and cook for the school's pre-Christmas stall, so when I cooked some puddings for us, I threw a couple extra in the pot. After that I had mums coming up and asking me to cook more. For the next two or three years I'd cook a few extra, then it got up to a dozen, and more and more people were asking. So I said to the mums the following year, why don't we make a bit of a fundraiser and actually sell them. We used our kitchen and ended up making 180 puddings. The school made two thousand dollars. It went from a little project to a big project very quickly.

The following winter we got lots of phone-calls from people who'd heard about us and wanted puddings for Christmas, and there were other people who'd bought two the year before and now wanted four. It was obvious that, if the school went ahead, it was going to get quite big. But the parents couldn't cope with the thought of that and so they backed off.

Neil and I thought, well, if people want this product and the school isn't interested — and it all originated here, anyway, using my recipe — maybe we should do it ourselves. That's when we did a bit of number-crunching, read a few Acts and talked to the Health Department people. It all came up in the positive.

My background was in nursing, not in marketing, but it's amazing how I was able to use my nursing skills to establish the business — quality control and the science behind how things work, and why you should do things in certain ways.

'We made 1,500 puddings in six weeks'

The first Christmas was mind-boggling. We made 1,500 puddings in six weeks, and it was all through word of mouth. Friends came to help. A business also contacted us wanting us to make hampers featuring our puddings. Anyone can go to the supermarket, and buy a packet of nuts and put together a hamper, but we decided we wanted to incorporate products and ingredients from the greater green triangle as much as we possibly could — which is how we came up with the concept we're now working with. The greater green triangle stretches from Kingston in South Australia to Camperdown and Horsham in Victoria.

After that first Christmas, Neil was diagnosed with narcolepsy, and he clearly wasn't managing the farm duties as well as he needed to. It was then that we thought about really developing the business, because there was clearly a strong demand out there. We figured it would bring in extra income as well as give Neil an opportunity to maintain his self-esteem.

Because we'd been looking at the green triangle for our hampers, we then starting looking at what products *weren't* being made in the area. No-one was curing the locally grown fruit or processing the yabby meat from the numerous yabby farms in the area. The fruit gets taken to the city market, and if it isn't sold there it gets sold off cheap to get rid of it — or it won't even get there because it's spoiled. And that's it. The income's gone. One man I knew used to feed his quail eggs to the dog because he couldn't sell them.

'We want to encourage local people'

So what we started thinking about value-adding for local products. We add value to our products by giving them a long shelf life — we can ship them all over the world because they keep at room temperature. We use our own milk in our puddings, and butter from the factory we supply. But we also try to value-add other local products. We use locally grown fruit and berries and, when we're in full production, we use anything up to 100 dozen eggs a day from the local egg-farmer. Now we're also curing fruit and pickling yabbies.

EMPLOYING APPRENTICES AND TRAINEES

If you are looking for an apprentice or trainee, there are a number of job network members who specialise in this area. These organisations work with you to develop a training program that is right for your business. They will assist you to obtain Commonwealth government training incentives and other kinds of financial assistance you are entitled to receive.

For the names of job network members in your area, and further information about them, contact Centrelink or your local TAFE college.

We want to encourage local people to get on the bandwagon. There's an opportunity for people to actually make some money.

Our local council has been wonderful. When we started, it was prior to the amalgamations. The environmental health officer at the old council helped us get registered and get all the information, but her underlying attitude was: 'You're mad! How are you going to make money out of plum puddings?' Then when the new Glenelg Shire Council came in, with a little bit more vision, they were just wonderful. By this stage we had export interest, and they could see the long-term picture and how it would benefit the region as a whole. They helped us get our planning permits organised and our extensions done. The chief commissioner of the shire actually opened the building for us, and that was a bit of buzz.

At the state level there's a lot of rhetoric about small business, but at the moment I can honestly say that there's nothing out there that helps us. It's all just words. Everything we've done we've funded ourselves. We think there's a lot of merit in what's happening here, and there should be support available to help others get established.

'We don't get a lot of support from the locals'

The family are all interested in what we're doing but we don't get a lot of support from the locals. We're known as 'the pudding factory'; no-one's got any idea of what else we do. It's been a surprise, I suppose, because I believe I'm a giver, not a taker. Before I got involved in my own business I put a lot into the community, and this is the first time we've actually put all our energy into something for ourselves. It's funny, because one of the things I really like about this business is the fact that we've been able to generate employment and income for other local people. That's my way of putting it back in. I'm not just a capitalist pig who's taking it out, but that sort of attitude does come through.

One of the benefits of this business, though, is that we've created a new group of friends, as well as developing new skills. We've met some wonderful people outside of the area who we'd never have met if we'd just stayed farming here. I suppose we've changed, and some of the people we used to be close to haven't changed with us. It's not that they've dropped off; you just don't have the same things in common.

There's a lot of vision for Country Cauldron. Because it's primarily a Christmas-based industry, we have a high cash-flow for November and December, and then a low cash-flow for the rest of the year. The new products we're bringing on, like the pickled yabbies and the cured fruit, are deliberately aimed at levelling out the bumps.

We're doing a promotion in England soon with the Australian Chamber of Manufactures. They chose our product and asked us to participate, and they've assisted us in getting some export-access funding — which is the first time we've received any funding from anybody.

We've also put in a very small waste-management system which is expanding by the hour as the worms multiply. Dale, who's now twelve years old, is responsible for that. We hope to sell the worms eventually.

'We have such wonderful resources here'

So there's lots of positives. The only thing that might hold us back will be financial constraints. Time will tell. I personally want to really push the concept of a regional perspective this year, because I think that we have such wonderful resources here that no-one knows exist.

The other thing we want to push is contract manufacturing. I've created five recipes for one company on the condition that the product is packaged here. They send the packaging to me and our staff do all the work, so we're bringing employment back into this region as opposed to just baking a cake and selling it. That's the sort of thing we really want to capitalise on, because there's opportunity for our kids — as in, our community — to have that employment.

Country Cauldron has probably put a burden on the farm in the sense that if we'd stayed farming and I'd worked three days a week we would be financially better off. We're ploughing everything into Country Cauldron, because we have this vision of what it can be.

From the bank's perspective, they see that as a risk and would probably prefer us to just keep on with the dairy. We needed to borrow $30,000 to do capital works for Country Cauldron, and were knocked back three times. If we'd gone in and asked for $30,000 to buy a new tractor they would have given it to us. They told us that.

The dairy farm as it stands is quite stable, because dairying is on an up at the moment. But there's all these cycles that you go through on a farm, and if you just stay as you are, you take the risk of being caught on a down-cycle. You're totally in the hands of the dairy industry, and you can either let them control your lives or you can diversify — which is what we've done. In the long term, the whole idea is that Country Cauldron should smooth out the bumps.

ENTERPRISE INFORMATION

Name of enterprise
Country Cauldron

Address
Nelsons Road
Winnap VIC 3304
Phone (03) 5528 1355

Type of farm
Dairy

Year established
1994

Product
Gourmet food products

District
West Portland/Dartmoor

Numbers employed
Four

Market
Gourmet food stores and delicatessens;
gift shops; wineries

Annual turnover
$100,000—$200,000

YARROW
Fax

15 Sue Jewson
Herb growing

Sue Jewson is thought to be the first person in the dry-land Mallee to have made a business out of growing herbs. 'Mystic Herbs' indeed: with a hungry marketplace and a potential profit of up to $20,000 per acre, Sue has tapped into one of Australia's burgeoning new industries.

I grew up on a farm over at Meatian, which is only 26 kilometres from Beauchamp. I worked in Swan Hill, then married a farmer and moved here. I guess I haven't moved far, but the Mallee is beautiful country and I believe it has great potential.

I started studying horticulture through Melbourne University to check out why things did what they did on the farm. We spend so much money on agriculture; I thought I should learn for myself. I wanted a trial plot to try out plants that I thought might grow commercially out in the paddock. My studies helped me choose plants that suited our land rather than just what I wanted to grow.

It's been a natural progression to the herbs. I wanted to do something to earn money that didn't interfere with my farm and family life. My husband Leigh and I have two girls, thirteen and eleven, and I like to spend time with them after school. I also wanted something that fitted into our main harvest and the sowing, because I drive the truck and tractor to help out. Working off the farm doesn't suit me. We're 35 kilometres out of town, and that's a lot of fuel money.

'Most dried herbs are imported'

The idea of the herbs came to me when I went to a field day at Seymour. Some manufacturers got up and said that they cannot buy

Australian herbs; most dried herbs are imported. What they import has to go through irradiation to come into the country and is often up to three years old before it gets on to our supermarket shelves. I was horrified to think that they were importing something that we could grow right here. I thought to myself, I can do that. So I had the market before I had the product.

Then I read everything I could possibly find on herbs — there isn't much on Australian growing conditions. I wanted something that went with our environment, which is dry and hot, and didn't use much water. I was left with the tough, hardy, Mediterranean-type plants.

'I started trialing different crops'

Two years ago I started trialing different crops. Some I won't grow again because they haven't worked out; they're too labour-intensive, or they require more water than I'm prepared to give them. I've only got an acre-and-a-half in, because we are on limited water supplies and I have to fit in around our main farming activities. This is not an irrigation area, and the dams are only filled once a year.

I originally sourced some seed from Eden Seeds in Queensland. Lately I've sourced seed from America via the Internet. I direct-drill most of my herb seeds — probably because I'm lazy. I don't see the point in sowing seed into trays, transplanting them into pots and then transplanting those into the ground. I drip-feed them for eight hours, once a week. It's not a lot of water because it's going directly onto where the plant requires it.

Everyone told me I couldn't grow echinacea. I didn't believe them so I direct-drilled it with a special little seeder that I borrowed from my aunt. The next night I went to bed and read another magazine and found out that it should have been chilled for three weeks. Fortunately we had some frosts following and I had a 100 per cent strike rate. I was thrilled.

After I had raised the first lot of herbs, I said to Leigh that I needed some sort of grading machine that would shake and grade the herbs.

Three weeks later he'd made me a grading machine.

I dry the herbs down in the shearing shed. You shouldn't process the herbs until after they've been dried, or else they will lose all the essential oils. Then we've got to get them into a presentable manner for the market. Some people want it for tea, some want it for medicines; often I don't know what size to grade it into. Leigh scoured the country for an old chaff-cutter so that we could cut it to any size.

'There is a huge market'

I sell some herbs in small packets at two Swan Hill outlets, but I mainly sell into Melbourne in 25-kilogram bags. There is a huge market. I try to choose plants that I have three markets for — oil, culinary or medicinal — so that if I can't sell the product into one market I can try the other two. I'm afraid to get into bigger markets, though, in case I can't supply twelve months of the year. That would be a lot worse than not having a market at all.

The first year I sold quite a bit of wormwood — that was to the medicinal people — but they're not buying wormwood at the moment. Sage, rosemary and yarrow are in big demand. I cannot supply enough. There is a demand for feverfew, but I've still got some stored.

I've sort of kept the herbs in the bottom drawer, so to speak, as far as the local community is concerned. Before I started growing them, I imagined people who grew herbs to be alternate lifestylers coming out of the hills saying, 'Hey man, we're growing herbs'. The herb word covers so much. Last year, when we planted out five acres of coriander, we did have a few vehicles come and stop, but they were a bit afraid to ask what it was. I do not know of anyone else growing herbs in the Mallee with a view to making an income out of it.

There is one other herb-grower at Quambatook, and I work with him. If a buyer rings me and asks for 50 dried kilos of wormwood, say, I ring him and ask if he can help me fill the order. We bag it separately because it has to go through strict quarantine procedures.

I grow to organic methods, although I cannot be certified organic because I live in a mainstream farming community. To be certified

How to Qualify as an Organic Grower

The Commonwealth Department of Primary Industries and Energy (DPIE), through its Australian Quarantine Inspection Service, has accredited various organic organisations represented on the Organic Produce Advisory Committee (OPAC) as capable of carrying out farm certifications.

OPAC has established national organic standards which have been approved by the DPIE. These standards set minimum guidelines that must be followed in order to qualify as an organic grower.

What is required for a property to be certified?

All certification organisations have production standards that conform with minimum national standards established by OPAC and ratified by the DPIE. These standards list allowed and prohibited inputs, and describe management practices which are seen as being consistent with the standards. There must be a minimum period of two years' conversion before a farm is eligible for certification status.

The certification bodies have different standards and different certification costs, so it is up to the grower to decide through which organisation he or she wishes the land to be certified. The point to remember is that it must be one of the DPIE-approved organisations. It is also a good idea to check with any intended wholesalers or retailers of produce, as some insist on a particular certification.

How to apply for certification

Growers interested in obtaining organic certification should contact the relevant certification organisation and obtain a copy of their organic standards.

If the grower is happy that his or her management practices are close to or consistent with what is required for certification, an approach should be made to the respective certification organisation, requesting an application form. This should be completed and returned with the relevant payment. An inspector will then contact the grower to arrange a suitable inspection time.

Based on the inspector's report, the completion of a comprehensive farm questionnaire, and satisfactory results from the soil or tissue analysis, the certification body will advise the grower what level of certification status has been granted.

For a full list of organic growers' organisations and certifying bodies, see the 'Further Information' section at the back of this book.

an organic farmer you can't have aeroplanes spraying within three kilometres; I wouldn't ask that of my neighbours. We send soil samples to Queensland for testing, and they send back a slip with all the chemicals that are registered. I have got clean soil.

'Water is our big issue'

I can't see expansion with drip irrigation. We've had some really dry seasons in the last few years. I would love to expand, but water is our big issue. I have to choose plants that give me the highest return with the least water.

I do see expansion with the herbs that supply their own seed and which don't require too much water; with these I can plant them down, harvest the seed, and then plant a bigger plot the following year. All of my herbs are grown in small plots, but if I find one that is suitable for broad acre, it becomes part of the farm business and is dependent on rainfall like our other crops.

Importing seeds from the United States can be very expensive. I've just got a packet of astragalus seed from America which cost ten dollars for ten seeds. It's a type of vetch that they're using with people who have had chemotherapy and drug treatment to help build up their immune systems. Coriander we've got going now; aniseed is an option that I'm trialing this year; psyllium I've trialed this year; and we've got ammivisnaga, which is a main ingredient used in the production of asthma medicine.

The potential is there to find something else; it's just a matter of being able to source enough seed overseas to get a few acres in at one go. I go to health food shops and find out what's in the tablets they're selling the most of, and then go back to the resource books and work out which ones I can grow. Our libraries in the Mallee don't have much, so I use the Internet. It was put in for the children, but I now use it quite a bit myself. There's also a Herb Growers Association meeting that I go to.

I haven't ever received information from any government departments or anything, but I really haven't gone to find it. I did have some interesting mail turn up through the shire, though. I think it must be from when I registered my business name, 'Mystic Herbs', because I haven't advertised.

'There's huge potential'

This is the end of my second season with herbs. Last year I wouldn't have made my costs, but this year has been good; I have made more than I expected to. They say you should be able to make $20,000 an acre from herbs, which I thought was totally wrong. But I honestly believe I can make that now, after growing just six rows of yarrow which brought in $400. I'm happy to be able to pay for the children's school fees and grocery money. I find that an achievable aim at this stage.

The biggest money is made by the people who get into a new industry early and then get out of it. Australia is one of the biggest users of natural therapies in the world, and to think that all the herbs available here are up to three years old and have been through irradiation treatment! There's huge potential for herb-growers in Australia.

ENTERPRISE INFORMATION

Name
Mystic Herbs

Address
RSD 445
Kerang VIC 3579
Phone (03) 5457 9295

Type of farm
Grain

Year established
1995

Product
Dried herbs

District
Beauchamp

Numbers employed
Self

Market
Local businesses;
Melbourne & Torquay wholesalers; end users

Annual turnover
$8,000- $10,000

16 Jan Maude
Alpaca knitwear manufacture

Jan Maude and her husband Ron switched to angora goats when the fruit industry took a downturn in 1975. Seven years later, Jan was launching her first line of mohair knitwear. Now she is at the forefront of the alpaca knitwear industry, and is helping to develop new technology to process the exotic fibre.

I come from Caulfield in Melbourne. I left the city when I was 21, and came to Shepparton to work. I met my husband Ron at the local dance, as country people do, and we were married within six months.

We came to live out here, next to Ron's father. At that stage we grew fruit. Ron had been on this property since his childhood, and he and his brother had planted 35 acres of fruit trees. The trees had only just started to mature when things became difficult in the fruit industry. We went through the struggle that all small fruit-growers had in the transition to bulk handling — the decrease in prices per bin, crop failures, all of those things. It was a really sad day for us when the bulldozers came to push the trees out, but with it came a certain sense of relief.

We took advantage of the Tree Pull Scheme which the government offered in 1975 — where farmers were compensated to remove fruit trees — and decided to invest in angora goats. They were at their peak at that stage, and it was an exciting time for the industry.

'It's a beautiful fibre'

I wanted to do something to subsidise the farm and I also wanted a challenge. I don't think I've ever been content just to work inside the house, so I looked around at what people were doing with mohair and started to think about value-adding. There were cottage industry-style garments, but not much being done at the boutique level. It's a beautiful fibre, light and warm with a wonderful lustre, and I thought I could do something with it to bring it up-market.

I didn't have much confidence in my own design ability, so I bought some commercial patterns and adapted them. Sourcing the yarn was difficult. I tried to get advice from people who were using it, but they were no help at all; they felt threatened by what I was trying to do. Fortunately I found Woolshed Yarns in Broadford who were processing a mohair wool blend. I also sourced brushed mohair from England, because it isn't processed here.

I started off with half-a-dozen people hand-knitting, and we produced 20 or 30 garments. Our launch was on April Fools Day, about seventeen years ago, at a garden party in Shepparton to raise money for the Spastic Society. We used our teenage kids as models. People bought the garments, so I was able to buy more wool.

We gradually employed more knitters from as far afield as Cobram, Bendigo, Albury and Echuca. We built a reputation, and people started to come back to us again and again.

I had no idea then what assistance was available to help people start in a small business. I just gave it my best shot.

'He said, "I'll buy everything you've got" '

Then I had a phone call from a man in Canberra who had a duty-free outlet. He said, in a gruff sort of voice, 'I believe you do knitwear'. I said, 'That's right, I could send you a video of our last parade'. And he said, 'I don't want a video, I want jumpers. Send me twenty'.

So I did. I got a cheque from him within a week-and-a-half. Then he wanted twenty more, and I was thinking, wow, this is wonderful. About three weeks later a Benz pulled up in the yard, and he walked in and said, 'I'll buy everything you've got in the shop, and you just keep sending them'. I built up my knitters until we had 170, and I was sending off fifteen or 20 garments each week to the one outlet.

Then everything came to an abrupt halt. The Canberra shop was servicing Japanese tourists, so the airline pilots dispute, followed by the death of the Japanese Emperor, meant all the people with money were staying home. I went up to Sydney and slogged it out on foot around all the duty-free outlets, but the blinds came down. Everyone was in the same situation.

We had adapted the size of the garments because the Japanese have shorter arms, and all of a sudden I was left with 170 of these jumpers. I had to pay the knitters, of course, and I'd already paid for my yarn. I'd actually borrowed to process the amount of fibre that was needed to keep the whole thing going. So I was left really financially devastated. The biggest lesson for me was not to put all my eggs in one basket.

'I think it's one of Victoria's best-kept secrets'

I started to try and think a bit laterally. We had a showroom on the property that we used to open on the weekends throughout winter, until we became busy with the Canberra outlet. I knew I'd have to go back to that.

I was lying in bed one night thinking of ways to get tourists here, when I thought, the first thing is that they have to know we are here. The tourist information centre was south of Shepparton on the Goulburn Valley Highway. We're on the east-west route, so I thought, we really need another tourist information centre here on the Midland Highway.

I love the Goulburn Valley — I think it's one of Victoria's best-kept secrets. The climate is good; it's green all year round because we have irrigation; we have big skies because the country's so flat; and the sunsets are magic. It's also considered to be the food bowl of Australia. I thought, why don't we do something that showcases what's available in the Goulburn Valley, including my knitwear?

So I went to our local government, the Shire of Rodney, and said, 'What do you do if you've got a good idea?' They gave me $150 to find a planner who would draw up what I had in mind. When I took it to the council they embraced it.

Not long after, the position of economic development officer at the Shire came up, and I applied and got the job. While I was working there, the tourist centre project actually picked up $700,000 in funding from the capital works program Keating brought in just before the elections. I think we called it the 'pennies from heaven' grant! I was appointed project manager for the construction.

The concept itself was very exciting on the tourism side. I was able to draw together a really broad range of farm enterprises that included Italian cheese-making, jam-making, a plant nursery, my knitwear, historical houses, and what have you. The Yooralla Society of Victoria ran the restaurant part of the project. We had two programs for unemployed Aboriginal people who designed all the landscaping for the site. I felt this was important because we ended up building on a site that was significant to the local indigenous people. Through it I learned a lot about Aboriginal people and their feel for the earth. Later, we built two new modules for small business, so there are now three buildings on that site. In 1993 it was awarded the best new tourism project by Tourism Victoria.

That project made me realise that you need to get a group of people together in a network to have any real power. Those people couldn't afford to individually promote themselves, but with tourism grants we were able to get them out into the marketplace.

The knitwear was put on the back burner while all that was happening, but when local government amalgamated I left that position to focus again on the business.

'The alpacas are absolutely beautiful'

Around the time that I was working with local government, our eldest son and his wife left to run an alpaca quarantine farm in Darfield, New Zealand. It was suggested that I do some knitwear in alpaca that could be sold from that farm. I loved the name Darfield, so I decided to do a new label — 'Darfield Alpaca'. I also thought it would be good marketing to have a local label.

As it happened, some of the animals were poisoned. The farm was sold, and our son came home and ran an alpaca breeding service from our property. However, I kept the 'Darfield' label and went on with it from there. We subsequently had 120 quarantined alpacas on our property that had come from South America via New Zealand. At that time, there was no protocol for alpacas to come directly to Australia.

The alpacas are absolutely beautiful. They've been domesticated for 6,000 years and they're very gentle. They have eye-lashes to die for and their coat is so soft. They talk to each other, which is a really comforting sound. They're also quite easy to farm; the worm problem is nothing like it is with sheep. And they're wonderfully hardy and very intelligent.

To shear alpacas you have to lie them down on their side and put their feet into a harness, front and back, which is then tied to a post. One side is shorn, then you turn them and shear the other side.

The difference between angora and alpaca fibre is interesting. With these exotic fibres, we've had to learn that we can't treat everything as though it's wool. Even with scouring techniques; alpacas have an open fleece and love to roll in the dirt, so you have to clean them in a different way. The same with dying and classing — you can't use the same techniques for all fibres.

'It's fun finding the markets'

I'm a strong believer that you haven't got a successful product unless it stands up in the mainstream market. With the alpaca knitwear, I do direct marketing exhibitions that have no connection with the alpaca industry. It's fun finding the markets. The market we have in Canberra came through word of mouth. They hold an annual expo up there for designers who manufacture their own garments. It's aimed at the upper end of the market. The person who runs it heard we had a quality product and he rang me; we've been with him now for five years.

We've only done badly at one exhibition. We went to the heartland of wool-growing in Victoria, and were frozen out because we were alpaca. I couldn't believe how blinkered the locals were about it. They didn't even want to feel it. I've never had that happen before. However, I guess it's understandable considering what's happened to wool over recent years.

I feel we should find a market in Sydney, but each knitter takes at least a week to do one garment. I've only ten knitters on the go at the moment. We've changed from hand-knitting to machine-knitting, because it used to take three weeks to knit a garment, and with the machine it takes only one. Also, you have a better-quality garment with the machine because the tension doesn't change.

I still advertise for hand-knitters who we then train onto the machines, because they have the patience for the finish that we need. Then I pay them the same as a hand-knitter, so they can earn in one week what they were earning in three. It's hard to pay knitters what they are worth. I always feel embarrassed about it, because I feel it isn't enough, but we are paying well above the average. I buy the machine and set it up in the home of the knitter, and we pay to train them.

My designer, Joanne, who's very talented, trains the knitters. She originally came to me to do hand-knitting. Then she came back with a jumper one day and said, 'I don't know what you'll think, but I thought this was a bit dull, so I did something to it'. She'd hand-embroidered on it, and it was just brilliant. We are good friends, and I value her immensely.

'I believe the future lies with finer knits'

Sourcing Australian alpaca fibre became a reality for me with the establishment of the Alpaca Co-operative. The co-operative enables breeder members to pool their fibre for processing into value-added product. The fibre is classed at the Gordon Institute of TAFE in Geelong, which ensures a quality product because the standard of classing is consistently high.

I'm currently a director of the Alpaca Co-operative, responsible for product development. A lot of time has been spent on research and development, but we're now getting results and putting in place a sustainable range of quality Australian alpaca products.

One of the problems faced by small business, especially in the fibre industry, is the lack of resources for small-scale production. Being a small manufacturer myself — as well as looking at the needs of the Alpaca Co-operative — it became increasingly obvious that there was a crying need for a fibre-processing company that would offer a service to small operators.

I have been involved in the establishment of just such a processing plant: Elite Fibre Pty Ltd. The company is located in Geelong, and is specifically aimed at small specialist runs. Elite handles all the luxury fibres — alpaca, cashmere, mohair — as well as the coloured sheep's wool and specialist wools required for value-adding in on-farm operations.

I believe the future lies with finer knits. Elite will be strongly focused on investing in the technology to allow its clients to compete with the best the world market offers. Australia arguably grows the best fibre in the world — we must also become very clever at value-adding.

'The potential for the alpaca industry is far greater'

I also do some consulting work for an alpaca importer who has built his own quarantine station on the Pacific island of Nuie. Recently there were 600 Peruvian alpacas on the island which are now in Australia. They will improve the gene pool significantly, because up until now the majority have come from Chile where they are peasant-bred. The Peruvians are line-bred, ensuring an improved gene pool for Australian breeders.

The potential for the alpaca industry is far greater than it is for ostriches or ever was for angora goats. With the goats, there was nothing done about marketing until it was too late. In the end, people were buying embryos for $1,000 each, and by the time the kids were on the ground they were worth five dollars. Suddenly the whole market was flooded. The Alpaca Co-operative is talking about value-adding and marketing now, while there's only about 16,000 alpacas in Australia. At the moment the technology doesn't exist in Australia to artificially inseminate alpacas. They have a gestation period of eleven-and-a-half months, and they seldom have twins. So the numbers will not build up quickly.

'We've come a long way'

We still have the angora goats, but we only do alpaca knitwear now. When I look back at the first garments I made, I can see we've come a long way. But you have to feel your way and grow with it.

I really believe in the potential of the business. I will persist with the hand-crafted range, as this gives us a unique niche in the market. However, I can see that I will need to introduce a finer commercial range because the demand is there. I'll only do this providing I can offer the same quality and finish on which we have built our reputation. I won't compromise on this, and am currently working towards this goal.

We're selling very well at the moment. At the peak, with the mohair, the turnover was $200,000. The clawing back has been somewhat slow because there was already debt from the farm; we were supporting a young family and trying to build the business up. Now that we have educated our children and they have moved on with their lives, we feel free to take more commercial risks.

Over the past couple of years we've had orders all year round; we used to get a bit of a slump in the summer. I need to train more knitters, and I need more exposure. I'll do extra promotions as knitwear production increases. The momentum is building.

ENTERPRISE INFORMATION

Name
Camden Park Collection Pty Ltd

Address
310 Minchin Road
Tatura VIC 3616
Phone (03) 5829 0210

Type of farm
Angora goats

Year established
1975

Product
Alpaca knitwear

District
Tatura

Numbers employed
Nine

Market
Direct marketing via brochure;
some retail; direct to public at exhibitions

Annual turnover
Over $100,000

17 Shirley Mayfield
Woollen quilt-making

When Shirley Mayfield was diagnosed with multiple food and chemical allergies in 1984, she made herself a pure wool pillow to replace her polyester one. Out of such a simple act a business was born. 'Aussie Wool Quilts' is now an award-winning enterprise with licensed outlets all over the country, producing everything from boomerang pillows to sleeping bags.

I didn't grow up on a farm. I was a fringe dweller in the country areas. My father was a bit of a gypsy; he kept moving.

When I married a farmer in 1972, we owned a 100-acre dairy farm down in Warragul. Over the years we diversified into potato-growing and rearing vealers. We also sold off some of the land. When the marriage broke up in 1993 we had 40 acres left, and I bought it.

I also had a small group of coloured sheep on the farm that I'd been breeding since 1975 because I wanted to do hand-spinning. I started off with just two black sheep, which gradually became thirty. I've been a knitter since I was four years old; I was always good with my hands. Although I left school at fourteen, I worked as a sales assistant and believe these skills also helped me with the business.

'My specialist strongly advised that I stop using polyester'

In the early 1980s, I had some health problems — chronic headaches and fatigue. Eventually they found out it was chemical and food allergies, and my specialist strongly advised that I stop using polyester because it affected me badly. Of course, all the pillows were polyester. The specialist told me where to get a wool pillow, but they were $40 in 1984 and I couldn't afford that. So I went home, washed a bit of my spinning wool, and made a pillow. It worked. It wasn't brilliant, but it was better than the polyester.

When I went back to the clinic the other patients wanted me to make them pillows, too. So I bought a bit of carbonised wool and just loose-filled some pillows. [Carbonised wool has been treated to get rid of burrs, and comes out in small lumps.]

I sold my first pillows for $3 each, just over cost price, but after twelve to eighteen months they were going lumpy. I kept experimenting and, after five years of trial and error, I started using slivers [loose untwisted strands] of carded [combed] wool. Certain types of wool are very bouncy. The authorities weren't much help; a local wool-buyer put me on to it.

I was also making quilts by this stage. I had to develop my own techniques for making the pillows and quilts. Feathers don't cling; wool does. That's why loose-filling is no good. As well as using the carded wool, I worked out that the wool had to be laid out and stitched in a certain way to stop it going lumpy.

'It really took off in 1989'

I think I was the first to start making wool pillows and quilts as a cottage industry. At first I mainly passed them on to other allergy patients who were desperate for pure wool. Unfortunately, due to labelling laws, items can be marked 'pure wool' and still be 5 per cent synthetic.

It really took off in 1989. In one year, totally on my own, I made over 1,000 quilts and underlays and over 1,000 pillows. I was well into my fifties by this time, and I just about killed myself.

I was also selling them all myself, direct to the customer, both to keep the cost down and so that I could get feedback. With direct selling I could say, 'If you've got any problems, come and see me'. A shop will refund money, but often won't tell the manufacturer about the complaint.

My only sales outlets were the Red Hill craft market and direct from home. Because it built up gradually, there was a mushrooming effect. One person would tell their friends, then the friends would tell other friends.

I put on one girl in 1991. I didn't want to get too big, just keep it at one employee. Then I had the bright idea of licensing people to use my techniques and market them in their own areas.

Greg Cahill at the Department of Agriculture had been in touch with me about my nomination for the Commonwealth Bank Farm Entrepreneurial Awards. Then, as a result of that, Mac Booth [a Rural Enterprise Victoria facilitator] asked me to speak to a group of farmers at a meeting in Bealiba in 1992. They were coming from far and wide to discuss ideas for things to do with wool. I said, 'They'll have to take me as I am'. I'd never talked in public before.

There were a lot of 'big' people there, like David Fletcher Jones and the manager of the Horsham Wool Factory. I felt totally inadequate. But I got up and spoke, and mentioned my licensing idea. Within two months, my first licensees, Peter Lakey and his wife Kathy, were set up near Bendigo. Peter had met me at the meeting. It meant that they could stay on a fairly marginal farm.

'For most licensees, it's basically about value-adding'

All the licensees that have come on have been in difficult situations. Some joined during the drought in New South Wales. One lady in Mungindi started because she and her husband looked like losing their family farm. She is now selling an enormous amount of quilts, and has also taken on the licence for southern Queensland.

For most licensees, it's basically about value-adding on the farm. It's something they can do from home and still be tied in with the rural industry. It costs under $20,000 to set up, and that's pretty reasonable compared to most small businesses these days. We pass on any good ideas, and also buy the packaging in bulk — so small producers can just get a box or two from us and not have a large outlay.

If a national paper contacts us, we'll sometimes put in a group ad; we put one in the *Farm Journal* mail order catalogue last year. We don't do a lot of commercial advertising because it brings poor results. Most of the quilt sales come from word of mouth, brochures, and attendance at shows, field days, home shows, and markets.

'We can make anything people want'

The licensees have a choice of which items they make, the main ones being the quilts, pillows and under-blankets. Some haven't got room for making mattresses; the mattresses and sleeping bags are not big sellers, anyway. The basic pillows come in five thicknesses, to suit everybody. We make different styles of pillows as well, including the 'Boomerang', 'Continental' and 'European'.

One of the beauties of our product is that we can make anything people want. We've got many combinations of 'duo quilts' for when one person gets hot and the other gets cold. Often people come in and want extra length or extra width. Once I even made an under-blanket for an eight-sided water bed. That was a challenge.

Licensees also set their own prices because different areas have different costs; but we ask them to keep within a reasonable range of their neighbouring licensees. There's an allocated area for each licensee, but the lines aren't drawn with barbed wire. We believe very much in the ethics of a business, and over the years customers have come to trust our name.

By 1995 I had eight licensees; some have dropped out since and more have come on. Unfortunately I've been let down by a couple of people who took on licenses and then went out on their own, using my technology. I could take them to court, but I'm not rich enough. So they're getting away with it at the moment. The ones who left the group but changed the patterns were not so much of a disappointment.

Most of these people are not doing as well as the genuine licensees — maybe because if they weren't dealing straight with me, they're not going to be dealing straight with their customers. It's hard when you take somebody in good faith, and they turn round and kick you in the teeth.

Most of the licensees are farming people like myself. The ones we had problems with weren't farming people. I think people who grow up on a farm understand cause and effect more. If you don't put fertiliser on your paddock, or don't look after your animals, you don't get any returns. Farming people are more straight in their dealings as a rule. I still trust people, though; I think if you stop trusting people you're the one who loses.

Another reason the people who have let us down may not be doing so well, is that integrity is sewn into the product. I feel that's why the wool quilts got so popular. People get a good feeling from them, even if they don't know where they come from. I think the type of energy that goes into the product comes out for whoever uses it.

I have two ladies who turn up every couple of years to get some carded wool. They always insist that it has to be new wool that isn't too long off the sheep. They use it as a therapeutic aid for healing. They say that because wool is a living fibre, the goodness comes out of it and becomes a healing agent. It's an interesting concept and apparently it works for them.

'I'm basically just a farm person'

The first year that the business really took off [1989], I sold $6,000 worth of quilts and pillows. Then in our last year at Warragul [1996] — before we sold the farm and a license for Gippsland and Melbourne to a lovely couple from the Bendigo area — our own sales reached $318,000 and we were employing four workers. This figure did not include our licensees' sales.

All I'm looking for from the licensing system is some sort of superannuation for myself. The average charge for the licensees is $2,000 per year. It's really a networking fee, because we keep in contact with them and help sort out any problems. We are considering, at the moment, whether there is a way that the licensees can eventually take over the company.

I'm basically just a farm person, not a business person. I love my animals. That's why I've retired, really. And, as I'm 60 now, it's time anyway. My brain keeps ticking over and thinking up new ideas. I want to get back into my spinning. I do all my own knitting designs but, because the Warragul business grew so big, I haven't knitted for three years. All we did was organise, make quilts, and go to shows. In 1996 we spent 70 days at markets and shows, including nine-day stretches and eleven-hour days locked in Jeff's shed [the Melbourne Exhibition Centre], so that took quite a toll on my health again. We realised it was time to stop. Besides, I have now married the man I fell in love with when I was eighteen, and I am selfish enough to want to spend lots of time with him.

'Maybe New Zealand, America, England … '

Most inventions occur because somebody needs something. I needed a pure wool pillow, but I didn't think it would grow this big. It never ceases to amaze me.

Over the years we've won quite a few awards. Twice we've been finalists in the Victorian Small Business Awards. In 1996 we won the Fair Trading Award, and in 1997 I helped with the judging.

We'll continue to service the licensees, and I would like to set up others in areas that are not yet licensed. Perhaps not only in Australia, maybe New Zealand, America, England … We got so busy that we haven't been able to follow up on any overseas prospects yet, but hopefully now I've retired we'll be able to do some of those things.

ENTERPRISE INFORMATION

Name
Aussie Wool Quilts

Address
Lot 25 Cuthbertson Court
Darlimurla via Mirboo North VIC 3871
Phone (03) 5668 2356

Type of farm
Now sold; previously dairy and potato

Year established
1989

Product
Woollen pillows, quilts, under-blankets, etc

District
Gippsland

Numbers employed
Self and husband running licensing
and product development

Market
Australia-wide, apart from Western Australia
and Tasmania

Annual turnover
Whole group — approximately $1 million

APPLE & STRAWBERRY JAM
Plum Sauce
APPLE & STRAWBERRY JAM

18 Wendy Mesley
Cottage-craft retailing

Wendy Mesley is one of four women who run a joint-venture gift-and-craft shop in Swan Hill. Wendy works two days a week in the shop, which stocks her crafts, preserves, sweets, and cakes. Because she has made no capital outlay to join the group, any money Wendy makes from her products contributes directly to the family income.

I joined the group twelve months ago. I saw an advertisement in the local paper looking for people interested in joining; I knew one of the ladies who was leaving, so I rang her. She is also a rural woman, and she'd been in the shop for three years, but she wanted more time to spend with her family and on her farm commitments.

She really encouraged me to do it. She said that, as long as I was prepared to put stock in the shop, it was quite financially viable. With encouragement from Graeme and my children, I decided to have a go.

'I registered my kitchen'

When I joined, we were trying to work out what all the new members could contribute. As I enjoyed making preserves and cooking, I first looked into that area. People asked me what I would do if the preserves didn't sell, but I have plenty of family to share them with. That was one thing that didn't bother me.

I registered my kitchen, and then started making jams, chutneys, and sauces for the shop. Then I went on to do home-made biscuits, fruitcakes, and sweets. As it gets closer to Christmas, I'd like to go on to do more Christmas things.

Some of the health regulations are a little bit hard to cope with in a small enterprise such as this. My kitchen was assessed by a health inspector, and he provided a twelve-month registration. I was told that if I'd decided at the end of the twelve months to continue baking, I would need to make some changes to my kitchen. The twelve months passed, and I decided that I couldn't justify spending the few thousand dollars that it would have cost to alter my kitchen. The main costs would have been for a new drainage system to take the kitchen waste, a separate hand-wash basin, and floor coverings that continued up the kick board. Instead, I'm now paying a local tea-room to prepare the biscuits so I can maintain the supply of freshly cooked biscuits in our shop. I may also look at hiring a registered kitchen to use for the next season of preserves.

'We like the natural look'

My husband Graham does all the labels on the computer at home. We decide what we like, and he organises the labels and prints them out of an evening. We like the natural look, and we didn't want to go commercial because I wasn't sure how involved I was going to get.

Deciding on the jars was the next step in getting things under way. I used the Yellow Pages to find suppliers of jars, and then went to Melbourne in the ute to check out their stock. As each company had different shapes and sizes, my final decision was based mainly on price. I only use new jars, and they all match. I think that's important — presentation makes a huge difference.

I use all local products. Because Swan Hill is an irrigation area, I can get all my fruits here. I nearly missed the apricots last season, but I got them in time. Then there were plums, and I did a little bit of peach and nectarine. Whatever I could get hold of. I did a lot of chutneys at first, eight different varieties, but I've learned that isn't necessary. People just tend to head for the plain old chutneys and not worry about the unusual sorts.

I don't have to have a use-by date — that's up to me — but the longer you keep your products the worse they look, and the harder it is to sell them. So I aim to supply fresh stock each year. I'm hoping that most of my stock will be gone by the end of the year, so that when it comes around to the fruit and tomato season in January I can replace them.

I've been experimenting recently with some other things to sell at the shop. A local person makes little wooden tables, chairs, and dolls' houses, which I then buy, paint, and stencil. All of this can be done at home. I've also been embroidering baby blankets and baby wear.

'We all take turns'

It's fairly quiet on the farm from January to March, which is lucky because that's when the fruit comes in, and I also have to work two days a week at the shop. There are three other women and we all take turns.

We actively advertise by attending our local field days, being seen with our product. Doing craft is by far the best way to let people know about our shop. We can also bring someone else's product into the shop, as long as it's different from what we already have. It has to be high quality — we are very particular about that. We have a group discussion about any new product, to decide whether it is appropriate for the shop. We bring a sample into the back room where we can all look at it, and if just one person doesn't want it then it doesn't go in.

There are also some local product lines which we sell on consignment. We try to stock craft from local people wherever we can. I regularly attend markets for new ideas, and go to the craft and trade shows in Melbourne. Our range in the shop is mainly cottage country wood, folk art, hand-made dolls, teddy bears, cards, and soaps. We have just begun stocking Australian-made souvenirs.

Each group member has a code which is used to identify her stock in the shop. This code is recorded along with the sales. At the end of each week, the sales for each individual member are calculated, and the money is paid from a joint account.

Some days I may only get $40 or $50, but I enjoy the company of people coming and going. We all take in crafts, or whatever we're doing, and sit out in the back room where we can see the shop. It's never a waste of time. As summer comes along there's always dusting to be done. That's going to be interesting — it's a battle doing my own at home.

During the holidays we run classes for school-children. I do gingerbread-decorating and chocolate-making. Other classes include folk art, card-making, and floral work. I also do adult classes in wool embroidery, with the other ladies holding folk-art, cross-stitch, and doll-making lessons. We advertise in the local paper, and the classes are usually full. Weekend workshops are held in folk art with specialist teachers, and all folk-art requirements are sold in the shop.

'I'm quite happy with the way it's going'

I really don't want to be doing more with the shop than I am at the moment. I'm quite happy the way it's going, and I couldn't take more time off from the farm. When I started, I said it wouldn't cut into my family commitments, and I haven't allowed it to do that. I have three kids in Melbourne and quite a bit of family around. If they need me, they always come first.

A better financial reward from the shop would be good. One of the reasons I really like it, though, is that there is no outlay other than the rent and running costs. When I first looked into it, I didn't know whether I was going to make any money. But now I can see that the more I put into it, the better the reward.

ENTERPRISE INFORMATION

Name
Cuddlepie Country Crafts

Address
RSD 448
Kerang VIC 3579
Phone (03) 5457 9325

Year established
1997

Product
Local cottage crafts

Type of farm
Grain

District
Swan Hill

Numbers employed
Four — all members of business

Market
Local and tourist

Annual turnover
$16,000

19 Judy Morrison
Plant nursery management

Judy Morrison planted seeds at dawn, before driving 60 kilometres to her day job in Hamilton, in order to start up her own nursery. Now tourist buses come in spring to see her 900 varieties of rare roses, and she employs six locals to help her run her country-wide mail-order business.

I grew up on a farm near Bendigo. I attribute my love of gardening to my mother, who was a great vegie gardener. I can remember, as a kid, helping her plant pumpkin seeds, and the great pleasure of eating *my* pumpkins. That's always stuck. By the time I was sixteen I had my own garden, and people would give me plants, which I treasured.

It took Noel [Judy's husband] and me quite a while to save up enough money to buy a property of this size. We've got 1,300 acres here, and the nearest bread-and-milk store is 24 kilometres away at Balmoral. We have lived in several places. We lived in St Arnaud for two years and farmed there on a very small scale. Then we moved to Phillip Island, where we had a small cattle farm. We didn't think much of having gumboots at every door, all year round, so we decided to come back to the dry country.

'We sailed in by the seat of our pants'

We bought this property at the fall of the wool industry in 1990. We sailed in by the seat of our pants, which is the usual story. You pour in everything and hope that you're going to get it out. Of course, the first wool cheque was a long way down on what we'd anticipated in our projections. Therefore it meant we had to make some choices. We either sold up and got what we could, or we had to do something else. I decided we needed a man — my husband — on the place to do the heavy work and I, being the woman, could work away from the property.

I took a job as a pharmacy assistant in Hamilton, which was a 120-kilometre round-trip every day. I was going to work in the dark and getting home in the dark. I thought, blow this, I can't be bothered doing this.

So I started growing things and doing a couple of the Sunday markets. I found that people really enjoyed the plants I was growing, because they were plants that I liked. I started thinking I could possibly make a business out of it, but of course we had no cash to do anything.

'My garden was calling me'

I continued to work in Hamilton so that I could set money aside to buy hot-houses and pumps and watering systems. I didn't have time to start my garden properly, because I knew I had to support our livelihood. But my garden was calling me.

I managed to change my hours to start work at 10 o'clock in the morning, so I was getting up at six o'clock and propagating for two or three hours before I left for work. In between the farm work, my husband was constructing watering systems and setting up hot-houses, so that the place would function by itself while I was away.

That went on for about eighteen months until we were set up. Then slowly, through word of mouth, people learned that I was here and that I was producing a good-quality product that would survive in country gardens. But I was tired of getting up at four o'clock in the morning to go to Ballarat markets, two-and-a-half hours away. I thought, one day I'm going to wake up and find myself wrapped around a tree. I decided that I had to make the nursery succeed so that I could stay home.

I really started marketing myself, telling people what we had and what we did, until we had built up a steady clientele. Then I started the mail-order system. I kept gathering names and addresses from people, and asking if they wanted my catalogue. I would sit up till the other end of the day, and type out lists of what I'd grown and what was ready to go. Slowly the mail-order system took off. In 1993 I gave up my job in Hamilton.

'We grow over 900 varieties of rare roses'

The main part of our business is mail order. We have a contract with Australia Post. I had to specify that it go through Balmoral Post Office to help it stay open.

We now have 20 to 30 tourist coaches come through each spring and autumn. They tour the garden and the nursery to see all our rare plants. I have tried to create a niche industry; we grow products that other people don't. You can't walk into any nursery in Melbourne and see our plants.

We grow over 900 varieties of rare roses. Some of them date back to before Christ. It's really something, to think that the genetic material that was alive before Christ was born is still here now. We have a little rose that was found on Omar Khayyam's grave. The things we have here are very special.

One of the reasons you don't find a lot of our perennials in your average nursery is that they do not make pretty pot-plants, but given room in the garden they're beautiful. Another thing is that a lot of perennials take more than two years to flower, which for the normal industry is too long for them to gain a return. The retail nursery sells a pretty flowering product, and we sell a garden-worthy plant.

It took a lot of work to source material. I sent overseas for catalogues, and now we import seed from pretty well all over the world. We propagate plants from seed and try them here; if they survive our hot summers and severe winters then we feel they're fit for the Australian climate. A lot of the roses were already in Australia. When I get some spare time I go out hunting. It's a great pleasure, and you find some real treasures.

After two years with the business I found that I needed help. I thought, I really can't afford to employ anyone but I'm going to. It meant that I had to work harder and market more. That's how it's

Criteria for Acceptance by Nursery Associations

Information can be obtained from the Nursery Industry Association of Australia Ltd, Cambridge St, Epping NSW 2121 [Tel: (02) 9876 5200] or the nursery association in your state. Their address details are as follows:

New Wouth Wales	Nursery Industry Association of NSW PO Box 13 Rouse Hill NSW 2155 Telephone [02] 9679 1655
Northern Territory	Nursery Industry Association of the NT PO Box 3829 Winnellie NT 0821 Telephone [08] 8988 1351
Queensland	Queensland Nursery Industry Association PO Box 345 Salisbury QLD 4107 Telephone [07] 3277 7900
South Australia	Nursery and Landscape Industry Assocn of SA Inc 198 Greenhill Road Unley SA 5061 Telephone [08] 8300 0172
Tasmania	Nursery Industry Association of Tasmania PO Box 17 Launceston TAS 7250 Telephone [03] 6398 2474
Victoria	Nurserymen's Association of Victoria PO Box 431 Caulfield East VIC 3145 Telephone [03] 9576 0599
Western Australia	Nursery Industry Association of WA 251 Adelaide Terrace Perth WA 6000 Telephone [08] 9325 8252

been all the way through. Every time we've taken another step, I've had to extend myself a bit more to make it work. Everybody around here knows me, but I don't know them because I work so hard. Sometimes I don't get out the front gate for six weeks.

Jacinta has been with me for five years now, and her knowledge has grown incredibly. Neither of us have any horticulture training whatsoever; I source information through books and horticultural colleges. Many people in the industry seemed to think we were a bit of a backyarder, and they were reluctant to pass on their successes and 'how-to's. The doors were closed. I found propagation techniques were hard to source, so I mostly learned through trial and error. I've taught myself how to propagate bulbs, roses, and perennials. We've just recently been accepted into the Nursery Industry Association of Victoria, so now I'm able to source a lot more technical information and material which will help us improve our productivity.

I now have six employees — all local people with farming backgrounds. The business is fully computerised, and we have permanent office staff. We've just put on two new trainees; hopefully they're going to continue on and do their horticulture training through RMIT TAFE [Royal Melbourne Institute of Technology/ Technical and Further Education], with the help of myself.

'They just laughed at me'

When we started up we didn't receive funding or support from any government bodies. We've basically done everything on our own. That way I feel that I'm only answerable to myself. If I fail, I haven't got a great big debt to pay back. We own what we have. The other thing is, when you're involved with a lot of these government bodies the paperwork is just unbelievable, and I haven't got the time.

I have been involved with regional development boards and that sort of thing. I sat at one meeting, and was absolutely floored when one of the local farmers stood up and said, 'I don't think that any land should be subdivided into less than 100 acres, because no-one can make a living off it'.

I stood up and said, 'Listen, pal, come out here and I'll show you how to make $90,000 out of ten acres, if you're prepared to get on your knees'. They just laughed at me. That was six years ago, and I have done it. It was the same with the bank. I approached them and asked for a nominal amount. We had a very good credit rating and we didn't have a farm debt — we paid cash for this property. Even so, they really scoffed at the idea of starting a nursery out here. I thought, blow you, I'll do it by myself and I'll show you. And we did. Now they're very happy to do whatever we want.

My husband still runs the farm as a fine-wool sheep property, but we have reduced the stocking rate because the returns aren't there. My husband came to me one day and said, 'I think we should buy that mob of 400 sheep'. I said, 'Let's sit down and look at what the return on that money might be if we bought the sheep, or if I bought a few thousand roses and resold them within twelve months'. The roses won hands down.

'Always go chasing rainbows'

I'd really love the business to level out now, because I'd like to refine it. I am continually improving techniques and cutting costs. Our aim is to service all of Australia with a quality product we can guarantee. We have had inquiries from overseas, but as soon as I tell them that they have to satisfy quarantine regulations at the other end, it usually stops them.

I have advertised on the Internet, but didn't get any feedback from it. Perhaps it was a little bit too early. In twelve months time there might be an opening there for us.

Marketing has never been a problem for me. I grew up in a business background: my family had several businesses, and I have had several businesses before coming into this one. You just have to tell people what you've got and promote it. Success breeds success. Never tell anyone of your failures — that's always been my strategy. There's been some hard scrapes, but I've always looked on things with a positive attitude and never given up. Always go chasing rainbows, as my grandmother told me.

ENTERPRISE INFORMATION

Name
Mistydowns

Address
RMB 440
Balmoral VIC 3407
Phone (03) 5574 3261 or (03) 5574 3244

Type of farm
Sheep

Year established
1991

Product
Roses, bulbs, perennials

District
Balmoral

Numbers employed
Self, husband, four full-time employees and two part-time employees

Market
Australia-wide

Annual turnover
Over $300,000

20 Val Murphy
Potato promoting

Val Murphy is often called the 'Potato Lady'. After marrying a local spud farmer she became committed to promoting potatoes to the public. Since then she has collaborated on two potato cookbooks, demonstrated her potato recipes at shows and festivals across Victoria and New South Wales and, in the past two years, has set up and catered for tours of the local potato-growing area.

I grew up on a farm in Gippsland. I was a dairy farmer's daughter who married a spud farmer. I met Phil through the Victorian Young Farmers Movement.

I knew little about farming — except how to help rear the calves and do odd jobs — because I'd worked in town before marrying. So when I came here at the age of 21, I was very naive. But it's been a wonderful experience, and I'm grateful for it.

I've always liked cooking, and I've always liked gardening and growing my own vegies — that way you know what's in them. I believe my health is my wealth, and if I'm able to provide good nutritional food for my family, and those who pass through our doors, that's great.

'It grew from the Thorpdale Potato Festival'

The first potato recipe book was published in 1976. It grew from the Thorpdale Potato Festival, which is held every year in the middle of the potato harvest — Labour Day Monday in March. Part of the festival is a potato-cooking competition, and in 1976 I was one of the successful competitors. That year, three other local farm women and I compiled recipes from the competition to put into a book.

The cookbook helps the potato industry by showing people the versatility of potatoes. Since we first published it, we've sold an enormous number by mail order; it's gone all over Australia and overseas. Last year we updated, and in that year alone [1996] we sold 3,000 copies. All the proceeds go to community projects and charities. I handle all the sales, and when it gets too much for me I ask a few local people to help.

When it's count month at the post office, I try to get extra orders by writing a spiel for the *Herald Sun* cooking pages and things like that; it helps to keep our post office alive and well. We're a small town and we need our little post office.

'I began to get involved in cooking demonstrations'

As a result of the cookbook, I began to get involved in cooking demonstrations — firstly for local fund-raising groups and then gradually further afield. Fire-brigade auxiliaries, church groups and the like would put on a day, perhaps a soup and sandwich lunch, and ask me to come along to do a cooking demonstration with potatoes. This helped raise funds for their particular group while also getting the message across about potatoes.

Over the years I've done cooking demonstrations all over Victoria and New South Wales — from food and wine festivals to agricultural shows. I always make a variety of things to hand out for tastings, and I talk about the nutrition in potatoes, the best varieties, and how to buy them. Potatoes are no more fattening than apples, you know; it's how you cook them and serve them that counts.

This is a big potato-growing area, so the demonstrations help protect the livelihood of many local people. I do what I can to promote the product.

About three years ago the local Women on Farms group was interested in visiting our property. They visit all kinds of farming properties, and they wanted to see a potato farm. Usually we all take our own lunch, but this time I said, 'Would you like me to put on a potato lunch for you?' So I hosted the lunch in our garden, and it was such a success that other people wanted to come. That's how 'Potatoes: Paddock to Plate' originated.

'It's all been word of mouth'

I'm now getting groups from all over Victoria — especially people from Melbourne who like a day out in the country. They'll organise their own coach and ask me to host. I've never had to advertise; it's all been word of mouth.

On a typical tour, we start off with a jacket potato lunch; we eat the skin as well because that's where a lot of the nutrition is contained. We have yummy sour cream and coleslaw, and bacon and cheese on the potato, with a side salad. Then I do a little cooking demonstration to show them how to make potato sweets. I usually make potato after-dinner mints, and they all try those. Then I talk about the cookbook which also has lots of wonderful recipes in it, from potato coconut roughs to baked-potato chocolate cake, which cut down on fats, oils and eggs by replacing them with mashed potato.

After our potato lunch, we travel by coach around potato country. We don't actually go into farms any more because numbers have become so great, but we see a lot from the roads. I tell people about the area and how approximately 150 semi-trailers, loaded with

Tourist Buses

Often, tour companies are seeking places of interest they can include in their tours. So you should not feel bashful about making a telephone call or, better still, a visit to local tour operators or those visiting the area to let them know what your enterprise could offer to their clients. Also, contact your local shire to find out where you can leave information about your business.

Another valuable way in which to make your enterprise known is through service clubs in the area; they often host members from other areas and may be willing to pass out information to any visitors. Your state government tourist body should also be able to assist. Contacts for these can be found in telephone directories.

potatoes, leave here weekly between January and June, to travel to markets in Sydney, Brisbane and Melbourne.

Then we do a nature walk to the Narracan Waterfalls, to walk off our lunch, and visit the site of the world's tallest tree. After that we park our coach in Thorpdale for about an hour, and I tell them about the country crafts and souvenirs that are available from the post office and the newsagency, and the baker who sells potato bread. I always tell the baker beforehand how many people are coming — there's usually about 48 — so he is able to bake accordingly. He always sells out.

So my tours not only help the potato industry but also our little township. They have even helped the primary school and kindergarten who have set up fund-raising stalls when the coaches have come through. Local church ladies also help cater for our lunches, occasionally, as a money-raiser. I think it's really great that it's got all these spin-offs.

At the end of the day, the tourists have potato ice-cream back in our garden and I give them a bag of potatoes to take home. So they've got potatoes in their tummy, potatoes in their hand, and maybe a recipe leaflet or the cookbook.

Last season I had 22 coaches within a few months, and all of those people seem to have told other people about their great day out. This season I think I'm going to at least double, if not triple, that amount, because bookings are coming in fast.

'I believe in getting off my butt'

I haven't really asked for support from the local government. I've had to register my business, of course, because I'm serving food to the public, but I haven't asked for assistance. I believe in getting off my butt and doing it myself.

I've been able to spread the word about the tours, and our local tourism operators are always happy to put me on their computer as well. A lot of Probus Clubs come and visit. They have a monthly magazine that they distribute, and they just have to mention what I do to another club and I've got someone else ringing up.

The media have been good, too. I've had television crews come up here from the 'Sunday' program and 'The Great Outdoors'. I've even been to Channel Nine's 'What's Cooking' program and cooked something on camera while also promoting our festival. I'm publicity officer for the festival, and last year they awarded me the Potato Industry Award in recognition of my work.

We find that instead of applying for government grants, we're able to help ourselves by working together as a community and providing a fun-filled family festival. That to me is very rewarding.

I can definitely see the business expanding. I now have two different tours, so that once people have come on Tour One they can come back and do a different tour. Tour Two has a potato lunch at the Mirboo North Brewery and then we go on a different nature walk and look at a different part of the farming area. It's still all potatoes; we just do different things.

'I can see myself getting busier and busier'

Next, I'm going to work on expanding the visits to overnight stays in the local town, where people can see the other things the Latrobe Valley has to offer — and there are many.

I can see myself getting busier and busier. Hopefully in the future I'll be able to employ one or two people to host tours if I've got more than one booking on a given day. In fact, I've been told that I really should be training an understudy in case I'm not well. So I guess that's something I should be thinking about.

In addition to the tours, I'm currently co-ordinating rural courses for local farmers through SkillShare and Gippsland Employment Skills Training. Some of our recent courses include quality-assurance management, computer training, and setting up a farm business on computer. Now we're looking at starting a chemical usage course for farmers, then fork-lift operator and first aid courses.

I also belong to a lot of committees and do work for a number of charities. It's endless. I can't understand people saying they're bored or they've got nothing to do. There's heaps to do out there. It's just time, isn't it? If you get to the point where you think you've done it all, you won't grow. There's always something extra. If you'd told me five years ago what I'd be doing now, and how many people I'd be hosting, I would have said, 'You really are crazy'.

'It's great to be needed'

It's nice to be able to help the livelihood of farmers, who are out there on the tractors every day, by promoting their produce beyond the farm gate. And it's great to be needed; to have something to get up for in the morning; and to be meeting people who are interested in what you do — people who want to know where their 'bread and butter' comes from.

Last year I was given a real surprise when the local shire awarded me Citizen of the Year. I was very embarrassed about it. I don't ask for any recognition for what I do; my rewards come from meeting lovely people and being able to help our little town and the industry. But I do hold great pride in the award. We have three children, all with good professions, and we hope they will carry on the name in the district and be proud, too.

I can honestly say I've never found anything negative about what I do. I tell people, 'If you've got any complaints, please tell me; if you enjoyed the day, please tell somebody else'. I've never had any complaints. The only negative thing is that I get called the Potato Lady. With a name like Murphy, what else could I expect?

ENTERPRISE INFORMATION

Name
Potatoes: Paddock to Plate

Address
RMB 1310, Saviges Road
Thorpdale VIC 3835
Phone (03) 5634 6267

Type of farm
Potatoes, onions and sheep

Year established
1996

Product
One-day coach tours,
expanding to overnight stays in local area

District
Gippsland

Numbers employed
Self; currently seeking to employ second person

Market
Victoria-wide; mainly Melbourne and Gippsland

Annual turnover
Covers costs and outlays

21 Glenys Nalder
Photography

Glenys Nalder used her background in art, combined with her experience of daily life on a grain farm, to set up a business selling 'photocards' of rural industry. The cards are now sold at a number of tourist outlets in Swan Hill, and Glenys is looking towards her next project: a book.

I was born on a farm. My elder sister and brother went to boarding school for their secondary education, but I was educated locally.

I did very well at high school, particularly in Art. I began a fine arts degree at the Bendigo College of Advanced Education, but I decided to leave and take a job on after the first twelve months. I knew from friends that *The [Swan Hill] Guardian* newspaper was looking for a new commercial artist. The requirements of the job were basically what I had been doing over the last year of my Bendigo course, so I approached them and then they made me a job offer.

After marrying Danny I moved from Swan Hill to the farm at Ultima West. I later moved to our other farm at Meatian where I currently live and work. I left *The Guardian* because distance was a problem. I was painting watercolours at that stage and continued until after my third child was born, not long after shifting farms. I have always taken a lot of photographs. My eldest daughter and my husband love photography, too, and they pick the camera up regularly. Many of their photos are just as worthwhile as any of mine.

'Not quite a gift card but more than a postcard'

The idea for the business came from my sister-in-law, Sue. She went on a holiday to Tasmania and came home with a card which had an aerial shot on the front of a Tasmanian shoreline. We termed it a 'photocard' — not quite a gift card but more than a postcard. The chap who had produced it had put a sticker on the back that stated who he was. And that was the inspiration: how to improve on the idea. I thought the card should have some sort of identification; a title; area information; and a name and address on the back as a sort of advertising-cum-copyright thing. Swan Hill had no area or industry promotion at that time beyond the simple postcards with a choice of about four designs.

I'd just enrolled my two eldest children in boarding school, so the early part of 1997 was rather busy. With two children still at home, it wasn't until well into Term One of the school year that I began to get some time to myself. That's what I've always found over the years: there's never a lot of time for yourself. Then, in early May, I put the idea together and made my first tentative moves into business.

'I began the photocard line with fifteen designs'

Sue had suggested rural industry as a theme for the photocards. We had a new header, which was the pride of our machinery plant, and both Danny and I had taken a lot of photographs of it during the previous harvest. I began the photocard line with fifteen designs, including five header shots from both mine and Danny's photographs. I made up 150 cards, ten of each different design. I initially had problems with tape and glue — the basic card construction — but I quickly worked out the best way to make them. There was a lot of time involved in making the initial batch, and I did a lot of figure-crunching. It was then that I realised that 150 cards was not enough to launch any business, and that I wouldn't really make any reasonable profit unless I streamlined, cut costs, and improved my card-making technique.

But the biggest problem that I had to overcome at that time was the scope of the whole idea. If I was going to be serious about a photographic business, I would have to put some money into the venture. I began with $1,000. I needed to reduce the costs of card production so I went to my developer and talked figures. I then went around Swan Hill sourcing my other materials: finding the cheapest places for card, the best tape to buy bulk, this and that and the other thing. It all adds up. At the end of the day, I found that I could slash my initial cost of production by 50 per cent and then that was converted to potential profit.

I kicked ideas around with my network of friends on how to sell the cards. Sue made a tentative approach to one place in town, and they were first to stock my cards. But that was just a launching pad.

'I approached the biggest motel in town'

I thought that perhaps one of the best places in town to sell cards would be one of the major resort motels, so I approached the biggest motel in town and offered them exclusivity. I put a lot of thought into how to approach the whole situation, how to present my ideas, how to promote my photocards, and so on. It was a lot of work, but necessary to making the winter school holiday deadline and utilising the tourist season. Although the idea worked well initially, in the longer-term I found the motel complex was not the best avenue for distribution.

The $1,000 soon stretched to $2,000 and then $3,000 as the number of cards required grew, and the necessity of displaying them well became obvious. To stock something like photocards, and display them well, requires finance before you even begin to count sales. My biggest problem was that I chose the quietest time of the tourist year to start. Swan Hill is very quiet during the winter. Then, during August, the Post Office in Swan Hill approached me requesting the opportunity to stock my cards after initially having turned down my request for their support. They wanted to trial the idea for a period, but time has proved my photocards a success.

My husband is very supportive — he's open to the idea of promoting the industry we work in and the area we live in. I have also found continuing support in the community for my photocards as I am promoting Swan Hill. Both locals and tourist buy them.

Sustaining Agriculture and Rural Communities

In 1995 the Victoria Women's Council surveyed women working in the Victorian agricultural sector. The respondents indicated that agriculture is becoming a more complex business. It is subject to continuing fluctuations in local and world markets, which require farmers to be more market- than product-orientated. Agriculture is also influenced by policy changes at all levels of government. In addition to this, agriculture is affected by adverse seasonal conditions, often without warning. Women from the north-east region told of devastating floods, and those from the north-west were dealing with drought and a severe mouse plague.

Respondents said that farming requires a capacity to innovate and adapt rapidly to change. As business partners and agricultural workers, the skills of women are vital to this process. Farmers are now faced with the decision of farming as a business, as a lifestyle, or a blend of both. The romanticised view of farming must be pared back to an understanding that it has the pitfalls of any other job or business.

The work of women in agriculture has until recently been seen as secondary or *unproductive*. There are gaps in statistical data relating to women's work on farms. Respondents believed that women's work both on and off the farm is a vital factor in the survival and development of many family farms. Many need an off-farm income to ensure economic viability. This requires an approach to rural development which ensures the growth and availability of employment opportunities.

Women identified strong links between the viability and sustainability of agriculture and that of rural communities. This understanding extends to recognition of the interdependence of rural and urban centres, and the responsibilities that both carry in preserving natural resources and the environment. In summary, the survey participants viewed agriculture as a much broader operation than just production.

For more information about this survey, see Getting to the Heart of Victoria: working with women in rural and remote communities. *A report on the Victoria Women's Council rural consultations. Melbourne 1995*

My eldest sister has encouraged me to do business courses, but I am a firm believer that you don't always have to have a ticket in something to do well in life. If you have talent, it will show through. You could do course after course in photography, but you must have talent and artistic flair to photograph well.

I don't have any business training, but there's a certain amount of 'feel' for business required with any business venture, and a need to be flexible at all times. Perhaps courses can teach you how *not* to do something, or teach people who have no business experience at all, but I have been farming for nearly twenty years; that, combined with the expectation that you will make mistakes in whatever you do in life, has given me a good sense of business. Besides, doing everything by the book may mean that you miss new vistas opening up, the need to simply take a different tack at times, or a different approach to coping with the problems life presents.

'It's simply live and learn'

Listening to people and gauging feelings is also useful. I think the quickest way to learn is to understand how I could have done something better. Confidence builds with every new experience. It's simply live and learn. I present myself to people and businesses in a way that tells them I have confidence in myself and my ideas. I think you need to believe in what you are hoping to achieve.

It's been a long road. I vowed a few years ago that I would never have anything to do with computers, but now the computer is one of my greatest business assets. I use a publishing package which allows me to print my own information on my photocards, letterheads, fax sheets and envelopes. I also use a business program for my cashbook and record-keeping. I save money by not having stationery printed.

But it's one step at a time. I'm taking the photos; I'm making the products; I market them; I supply them; I make displays; I restock displays; I handle the finances. I run the business by myself. I'm at the point where I'm ready to expand it, but I have to finish the things I've started before I take the next step.

I've had one or two people offer to help with the marketing side of the business, but delegation is not my strong suit. I like to know what is happening. It's not that they can't sell my product right, it's just that I need to know all the details. Perhaps I would consider employing people in the production side of the photocards, but I think there is a limited market for them at this stage.

I've had requests for gallery-type shots from people, so I can imagine expanding into that area. But my personal idea of expansion would be to do my own book. I always thought I'd write a book, because I do creative writing, but a book of photographs is something I would like to do.

I've got the opportunity right now to do what I want with the business. Our farm is subsidising me at the moment, but the idea is that the business will start to produce enough income not only to support itself but to help us out. The profit will go back into the business at first, because my camera is my husband's and it's a dinosaur. But once I've bought some equipment, if there's any money available from the business, I'd like to do some travelling.

ENTERPRISE INFORMATION

Name
Nalder Photographics

Address
'Mallee Downs'
Meatian RSD
Swan Hill VIC 3585
Phone (03) 5039 9554

Type of farm
Grain

Year established
1997

Product
Photocards, press photos, sports, rural

District
Victorian Mallee

Numbers employed
Self, with assistance from husband and daughter

Market
Rural newspapers, schools, rural interest groups; sports clubs

Annual turnover
So far has covered expenses and start-up costs

22 Jeanette Basset-Pearse
Host farming

When farming hit a downturn seven years ago, Jeanette Basset-Pearse and her husband Laurie bought an old church, moved it to their property, and converted it into a guesthouse. Jeanette now manages three popular guesthouses on their host farm in Donald, inviting her visitors to join in the farming activities, meals, and occasional late-night card games.

I've lived on this farm all my life. When I was nineteen I got married and shifted into town for a few years, then we shifted back out to the farm. My husband Laurie has been farming for Dad for about seventeen years now, and Laurie and I have been doing the host farm for seven years.

We first looked at doing host farming a few years before that when my parents were living with us, but we weren't really that encouraged after speaking to a woman who was running a host farm. Then my parents shifted to Lakes Entrance, and farming was on a downturn, so we decided to go ahead with our new venture. Even though Laurie does carpentry and farming, we only had about 1,200 acres, and we really needed to diversify just to bring in the extra income so we could remain on the farm.

'We came up with the concept of the host farm'

We originally came up with the concept of the host farm because of Laurie's sister. She worked as a nanny to some missionaries, and we thought it'd be a good place for the missionaries to come when they needed some 'R & R' [rest and relaxation]. They still use it, but now they only make up a small proportion of the guests.

The beauty of the idea was that I could stay on the farm and help Laurie with the truck-driving and general farm work at the same time as running the host farm. Graham Harris, the REV [Rural Enterprise Victoria] facilitator at that time, helped us with the business side of it and told us about the NEIS course. To do the NEIS course I had to register as unemployed; I didn't find that easy at all, but the course was good.

We had a self-contained flat that was close to the house, so we initially used that to see how we liked the idea. We did enjoy it but it didn't give us much privacy, so we bought the first church in Dimboola and moved it onto the property. We later bought a second church from Jeffcott, a small parish near Donald, and moved that onto the property as well. It was the same church that Laurie went to as a boy. We now have all three available for visitors.

We completely restored and renovated both churches; one of the churches has a rabbit theme and the other has bears. The balustrade we put into the Peppertree church was actually built out of the altar rails from the church where we were married. Each year, for our anniversary, we go and stand there to reaffirm our vows. All the decorating was a joint effort, but it's my baby as far as running the business and marketing goes.

'We have repeat bookings'

We have had a lot of support from different people. One friend said to me when we first thought about it, 'Who would ever want to come to Donald?' But it has been successful and it's still growing. We have repeat bookings. People are very interested in coming from the city to join in family life on the farm: to see how country people live, and the way to shear a sheep and sow grain and shift stock and ride horses. We try to make our farm guests feel like they're part of the family, and to do that you have to have a lot of interaction with them. I nearly always give them Devonshire tea before they leave, and we have meals sometimes and socialise.

At Last: Studying Women's Contribution to Farming

In 1998 the first-ever major Australian study of women's economic contribution to farming was published. The report found that Australian rural women make a substantial contribution to farm output and to the maintenance of rural communities. Their contribution is often undervalued because much of their work is 'invisible' — unpaid and simply expected as part of their roles as mothers and wives.

This invisibility has allowed policy-makers to largely ignore the effects of government agricultural policy on women and the potential contribution of half the people living and working in rural Australia. It has also led policy-makers to undervalue the economic and social environment for women in agriculture.

Estimating women's contribution is a first step towards policy change. By providing information on the relative and total involvement of women in agriculture and producing estimates of their potential, this project should contribute to change in the focus of policy-makers and should eventually improve outcomes for the whole of rural Australia.

As it happens, much of farm business is small business — especially when the vast majority of farm businesses employ very few people — and women's involvement in Australian small businesses of all types has been growing and prospering over recent years. The rate of small-business ownership by women is growing at approximately twice the rate of men, and women's survival-rate in small business is higher than men's. In addition, women are taking a lead in introducing both new enterprises and adding value to existing farm activities.

Women in agriculture provide paid and unpaid work in both the market and non-market sectors of the economy. The study project concerns women in agriculture and resource management, but there is an overlap with non-agricultural activities because many women who live on farms work off-farm in non-agricultural industries, and maintain much of the fabric of rural society.

For more information about this study, see Missed Opportunities: harnessing the potential of women in Australian agriculture. *Volume 1 and 2. Rural Industries Research and Development Corporation. Department of Primary Industries and Energy. Canberra 1998*

We've always loved having visitors, and really it was just an extension of doing that. We found it really hard to take money from people at first, because we felt guilty — it was something that we loved doing. But we've certainly learned since then that it takes a fair bit of work, and we need the income to survive on the farm.

It's tricky to know how far to go into the marketing, because it's expensive, and you've got your setting-up costs and the rest of it. We're very happy going through the Host Farm Association. They come and inspect the property and have a meal with you, and then they sell it to people. We pay about $500 in advertising and then, every booking they have, we pay 12 per cent commission. The visitors often come back to you directly, which saves paying the commission. That's the main way, and also through word of mouth.

'They're just like family friends'

We feel that we've got enough on our plates for now. We often play cards with our guests, and we might be up until two in the morning. Then we have to get up again at six o'clock, so there's not a lot of sleep during harvest time. But we find that interaction with people brings them back time and time again. They're just like family friends.

It's been a very worthwhile enterprise. I'm much happier staying on the farm than having to go out and get another job, because I can be with my family while I've got them. And it's been a real bonus for our two boys; they've made lots of friends out of it. Farm guests have even asked them to go back to stay with them in the city, and go to the tennis and things like that.

Since buying the Jeffcott church, the local people have also shown a lot of interest in what we're doing. That church was falling into disrepair, and many people were very, very pleased to see it being restored; they knew that Laurie would do a good job.

The other idea that we've got, perhaps down the track, is to give one of the churches to each of our children; but certainly not for some years yet.

ENTERPRISE INFORMATION

Name
Peppertree Cottage Host Farm

Address
'Peppertree Cottage'
Sheephills Road
Donald VIC 3480
Phone (03) 5498 6265

Type of farm
Grain, sheep, cattle

Year established
1991

Product
Host farm holidays

District
Donald

Numbers employed
Self, husband and children help out

Market
Mainly Melbourne and Adelaide

Annual turnover
Approx. $10,000

23 Christine Picken
Flower growing

Christine Picken went grape-picking and worked in a fish-and-chip shop and local timber mill in order to start up her own flower business on the family farm at Chetwynd. Now, after three year's 'hard slog' and the help of a local business course, the market for her flowers is expanding and she is enjoying the first real returns on her investment.

I was brought up on a dairy farm in a large family, and I always said I wouldn't marry a farmer. Of course I did. When I was a full-time mum I'd help out on the farm — being a rouseabout, feeding the shearers, and normal stuff like that. Then we diversified and got the emus and deer. People thought that was a bit strange. There's nobody else in this district who has diversified like that.

'I didn't know a lot about it'

Now I've got my own business doing cut flowers. I've been in it for three years this year [1997]. The idea first came from a neighbour who has since moved away. I decided to do it out of a combination of needing something else to make money and because my kids were grown up and I missed them. I didn't know a lot about it before I went into it.

The Warrnambool Co-op was going at that time, and they were after people to grow flowers. So I thought I'd have a go. I started off with two hundred, then down the track the Warrnambool Co-op shut. I thought, uh-oh, what am I going to do? Then I found out they had a business course running over in Hamilton — part of the New Enterprise Initiative Scheme — and I went with my daughter to check it out. I decided if I was going to do anything with the flowers I'd better do this course. During the course I learned that I had to look at where my markets were. If I grew these flowers, who would buy them? That's when I started really getting into the nitty-gritty of things.

The size of my market has gradually increased. I do a local run and I've recently expanded to Mount Gambier. I have a wholesaler in Sydney who buys my dried products, which I'm very pleased about. She found me; I didn't find her. It was through connections with the Warrnambool Co-op. My daughter, Cindy, in Mount Gambier, does most of my public relations work for the florists there. She's wonderful at that.

'It's been a hard slog'

John and the family have been very supportive, but no farm money has gone into the flowers. To get the flowers going, I've gone out and worked part-time in a fish-and-chip shop; I go grape-picking every year; I spent a year working as a 'casual' at the timber mill; and I generally do whatever I can find. Last year I went away grape-picking because I needed a cool-room to store the flowers in. The only thing the farm has contributed is the water and the land. It hasn't been easy, I can assure you. It's been a hard slog, because the men [Christine's husband and two sons] have very little to do with the flowers.

I found it very, very hard to get help from other people. You've really got to do your homework and find out where to ring to get help. You've got to be able to talk to other people in the industry and get information; to take time off to go to seminars and conferences run by the Department of Agriculture. Now that there's a few more seminars happening it's a little bit easier, but when I started there was nothing, and I didn't have anybody to turn to.

I prepared all the flower beds myself with a bit of help from John. I'd see some old posts and rails and think, mmmm, you won't miss them . . . I would make do with old cyclone wire left over from fencing, and other bits and pieces around the place.

Producing Cut Flowers

If you are thinking of getting into the cut-flower business, you should seek production advice from every source available — state departments of agriculture or primary industries, growers in your region, wholesalers, florists, industry associations and exporters.

Ideally, you need to consider many factors before committing yourself to such a venture. While this may sound daunting, you need to be well prepared. Some of the main factors include:

Capital

Acquiring suitable land is only a small proportion of the establishment costs. As well as securing adequate working capital, you need to be able to cover the costs of suitable buildings, machinery, equipment, water supply, irrigation, plant material, cool storage and post-harvest facilities.

Knowledge

The number of cut-flower producers is increasing each year. Many enter the industry without formal horticultural training or skills. However, possessing good management skills and using professional advice are essential if you are to succeed. You should have a good understanding of the most suitable varieties of cut flowers or foliage for your particular area; budgeting; production processes and production costs; and the best way of selling and marketing your products.

Transport

The cost and reliability of transporting cut flowers to market or point of sale has to be considered. For fresh flowers, it is preferable to use transport that has refrigerated cooling to 2-3 degrees Celsius.

Quality

To ensure maximum returns, it is essential to have a high-quality product suitable for the market. Many easily grown crops are over-supplied at peak periods, resulting in poor returns. Check out market niches, extended seasonality, new varieties, and so on, to enhance your chances of receiving good financial returns.

This information is based on material supplied by the Flower Export Council of Australia [FECA]. For more information, contact FECA and/or the Flower Industry Association Australia at 24-28 Collins St, Melbourne Vic 3000.

I dug the post-holes and nailed the rails on with my little girl, Jenna, after school. In the first patches I didn't leave enough room between the flowers. You learn from your mistakes.

I've experimented a lot. I've got a small hot-house where I can grow seedlings, and a worm farm to break down the manure. I use the worm castings mixed with sand and emu manure. It's very laborious work. I have a bit of a break in winter, but there's also a winter crop and I have to put the bulbs in then.

I put an awful lot of time into the flowers. I used to be a full-time mother — three hot meals a day — and now it's one hot meal a day. They get their own breakfast, and then we have a sandwich for lunch. Some weeks I put in seven days a week; other weeks, if John's got something else happening with the animals, it might be two days.

'The flowers are now starting to bring in a bit of money'

The flowers are now starting to bring in a bit of money, but it's taken three years to get a return on them. I've got quite a lot in now, but there's been a heavy outlay to start them up. The Misty Blue is my biggest seller. It's a tissue-cultured plant that I have sent down from Sydney.

With the extra money I can do a few little things that we haven't been able to do before. I've been able to make the house more comfortable. I do my shopping on the days I do flower deliveries, which kills two birds with one stone.

I would like to get big enough to employ somebody and get a van so I can transport the flowers around. I've got enough work to employ somebody, but not the money to do it. That's my long-term dream. Jenna spends a lot of time helping me strip and bunch the flowers. I guess she's my little employee.

A lot of the people around here haven't been over yet to see what I'm doing, but word is getting around now and they do talk about it. I'm getting recognised for what I'm doing — having a go.

The flowers are something that I can call my own. Something that *I've* been able to contribute.

ENTERPRISE INFORMATION

Name
Christine Picken Flowers

Address
RMB 1882
Casterton VIC 3311
Phone (03) 5583 1518

Type of farm
Sheep and deer

Year established
1994

Product
Flowers

District
Casterton

Numbers employed
Self

Market
Dried flowers sold in Sydney;
fresh flowers sold locally & regionally

Annual turnover
Approx. $5,000

24 Jenny Stuchberry
Pheasant farming

Jenny Stuchberry and her husband Cec started to look into the market for pheasant meat in Victoria after seeing pheasants in the wild in England. They now farm pheasants, partridge, guinea fowl, and Chinese Silky Bantams, which Jenny sells to restaurants in Melbourne and the local area. Jenny is now collecting exotic birds with a view to setting up a picnic spot for tourists on their property.

I grew up in the city. When I married Cec we lived in Western Australia for a while, but Cec was born and bred on a farm in Pakenham Upper, and he always wanted to get back on the land. We bought 30 acres and ran goats for about six years. Then we sold that and moved here.

We've only got a small plot here — five acres. My sister and her husband wanted some land, so they bought 700 acres; but my brother-in-law works in Melbourne, so Cec runs the property with the help of my sister. The shire allowed us to sub-divide five acres of it, and that's how we came to be here. I also work off-farm at the local community health centre three days a week.

'We had to learn about hatching'

Originally we saw the pheasants in the wild in England and talked about them for quite a few years. Then we decided, well, why not farm them? But we didn't know what the demand was.

I rang round, but nobody knew where I could get any information. It took me phone call after phone call. I rang the Department of Agriculture; nobody could help me. After I had literally made twelve phone calls and was about to give up, somebody gave me a phone number and said, 'Ring there'. We got onto a chap at Pakenham and went to see him. At first he seemed to be trying to put us off, but as the day went on he began to give us more information, and he's been very helpful since.

The first year we bought the eggs from him once a month. We'd buy 100 at a time and they'd come up by train. It was a big learning process. We bought small foam incubators, and had to learn about hatching and brooding; we had never done anything like this before. We'd had animals but we'd never had birds.

We got into silkies [Chinese Silky Bantams] after the pheasants because our daughter decided to give them to us as a Christmas present. I could have cursed her at the time, but now I enjoy them. If we had more of them we could sell to the Asian market in Melbourne, but we don't have enough. We have quite a few local people who buy them as pets for their children or for their garden. They don't wreck the garden like the larger chooks. We have orders for some, but we haven't started hatching them yet. We'd like to branch out and get a few more different silkies and have them as a speciality. I didn't realise until a few months ago how many there are.

'We are also trying guinea fowl'

We wanted to try a few different types of birds, as well as the pheasants, so we are also trying guinea fowl. People buy them as pets because they are good watchdogs and they stay around the yard, but we can't get a market with the restaurants. We also have partridge, which are very hard to rear. They're totally different to the pheasants. Each type of bird is farmed differently.

My family thought the birds were a good idea, but a lot of people said, 'Oh, not again'. When we went into goats they said the same thing. But we've had a lot of interest. We had people out here yesterday wanting to have a look at our set-up. It's quite a talking point, but I am so tired of hearing that awful saying, 'You're not a pheasant plucker but a pheasant plucker's son'.

Cec does the heavy work, like feeding and looking after the

incubators. I do the other side of the business, like marketing, the books, and publicity. I love the marketing side of it, meeting with prospective clients. When we decided to take on this business, I did the twelve-month small business course at TAFE, to help us set up and understand all the aspects of small business. I found this very interesting, especially the module on marketing.

'We had to find our own markets'

Peter from BARA [Business and Rural Assistance] in Orbost was very helpful about the financial side of it and marketing. He recommended that we look for a market before we actually started, rather than just getting stuck with birds that nobody wanted.

We had to find our own markets. I was ringing up restaurants, writing to restaurants. I've been down to Melbourne quite a few times calling on restaurants. There is a lot of foot-work involved in that. I was a bit dubious about getting in touch with the big restaurants, but a friend of mine suggested that I ring Stephanie Alexander. I thought she wouldn't be interested in a small concern like ours, but after a lot of pushing I eventually called and was very surprised when she said, 'Yes'. That gave me a lot of confidence to call other well-known restaurants. We've got three good restaurants at the moment, and that is all we can handle, but next season we hope to be able to supply a few more.

I have a friend who is a chef at TAFE, and she has been helping with the marketing and also developing recipes for the birds. We have done promotions at food fairs, but did not find they helped; they only cost us money. My friend Martja and another chef from TAFE have a program on the ABC radio once a month, on a Friday morning, and her recipes for our birds were featured on the program. This gave us publicity in an area where we weren't known, and we received a couple of small orders. Every little bit helps. Hopefully it will grow from there.

When we started out, we had to build up slowly, bit by bit. I went to VEDA [Victorian Eastern Development Association] but they didn't give us much help. Then Peter [from BARA] said to us, 'You should do a business plan, then take it to the accountant and see what he thinks'. This was before I started the small business course. I did that, and the accountant was so impressed that he said he would be able to organise financing if we had a problem; but we went to the bank, and they have been very good.

The fact that we've both got other guaranteed income makes a big difference; we wouldn't have been able to do it otherwise. But we're not rushing into things or taking big overdrafts that we can't cope with. I'm too scared. That's why we started off with little incubators and three or four pens. We've grown from that to what we've got now — twenty pens. Because we're only getting a return for six months of the year, if that, we have to wait until the money comes in before we do anything. A ton of feed is coming tomorrow for the breeders, so we've got to find $500 for that.

I do all the accounts on computer. Before I got the job with the community health centre, I'd been out of the work-force for a long time because of a back injury. I was told I was unemployable, so the

Using the Internet

The majority of government departments are on the Internet. The ABC also has a site for rural information and would be an excellent point to begin with. You can find this site at http://www.abc.net.au.rural/

The Victorian Rural Women's Network has a page which gives access to information and resources for rural women. Their website is at http://www.nre.vic.gov.au/commuty/rwn/rwn.htm.

These are only two of a myriad of sites which will act as good starting-points for information you need. For those who do not have access to a computer or the Internet, many schools and libraries allow you to book time on the 'net. TAFEs and many learning centres have courses on the use of the Internet. A number of rural communities have also established Telecentres in their areas. The effort it requires to access one of these is well worthwhile — not just for seeking information but by allowing you to communicate on a one-to-one level with other women on the land.

CES sent me to TAFE. I did twelve months there and couldn't get a job. I was also told my age would stop me getting a job. I went back to TAFE and did the Advanced Course of Office Administration. As well as getting me a job, the course has helped a lot with the business; I'd have been lost without it. I enjoy computers, and we've set up the same program that our accountants have, so that also saves us money.

'There is no big network'

I haven't got onto the Internet yet. I can't justify the cost of it at the minute. A friend of ours went through the Internet for us, but he couldn't find anything much on what we're doing. It was all American, and they have mainly game birds, not meat birds.

There's very little written information, either. The chap that helped us get started has written his own book, which was published at the end of 1996. That is probably the best resource book. He showed us different ways of keeping a record of our better birds, which is something that I've set up this year — toe-marking. He's got it down pat so that you can see at the end of the season which cage has been producing the best.

There is a Pheasant and Water Fowl Association, but it's mainly just people who are doing it for a hobby, not people who have gone into it as a business. There are two other small pheasant farms in the area. We have got to know them, and we swap information, but there is no big network. It would be good if there was. That's why we're going to Warracknabeal on the way back from our holidays, because there's a chap there with a big set-up. We're going to go and have a look and see how he's going. He says there's a high demand for these birds with the restaurants, so it will be interesting to see what he's done.

We want to get to a position where at least one of us can stop the off-farm work. If things go the way we want them to — if this business can more or less cover my wages next year — then perhaps the year after I'll be able to stop work and spend more time building up the business. I'm going to have to take a holiday soon so that I can go to Melbourne for a week to market the birds.

The birds we've got now are basically for the table. We'd like to set up a separate area and have exotic birds, pheasants and native birds. The long-term aim is to encourage tourists to come here. We'll have a picnic area and perhaps a coffee shop. Because it's a tourist area, and there's nothing else like it round here, hopefully it will attract a few people and perhaps make a bit of money, too.

I can imagine what it will all look like. I sit down and plan what we're going to have. By the time we've got our trees growing around the cages, that will take the bareness away and it will be quite attractive.

ENTERPRISE INFORMATION

Name
Nyrilla Pheasant Farm

Address
PO Box 727
Bairnsdale VIC 3875
Phone (03) 5157 7542

Type of farm
Cattle and grazing

Year established
1995

Product
Pheasants, guinea fowl, partridge,
Chinese Silky Bantams

District
Bairnsdale

Numbers employed
Self and husband

Market
Restaurants in Melbourne and local district

Annual turnover
Under $2,000

25 Judy Thomson
Plant nursery management

Moormurng — 'place of trees' — was the name Judy Thomson gave to the nursery she established on their family farm outside Bairnsdale. Judy does all the propagation, marketing and book-keeping for the business, which is now the family's primary source of income.

I grew up in Bairnsdale. My only experience of farming was on my father's property. As kids, we used to ride horses and sit in the back of the trailer with the hay. It was all a bit of fun. The property was an outside interest for my father — more of a hobby than an occupation — but one that he loved.

When I married Graeme we moved into his father's house. My father-in-law lives in Melbourne. The property consists of 1,350 acres and is predominantly used for fine wool production. Eleven acres of it was ours, and we soon built a house and a garden, and planted wind-breaks. That was when we became aware that there was no-one in the area selling native plants in tubes. The only native nursery sold all their plants in 6-inch pots, and you can imagine how costly it becomes when you are first starting out.

After I had my first child I was looking for an interest. We had also come to realise that farming was not returning enough income for a suitable standard of living. So we started growing trees and shrubs as a business, filling the need in the area.

'Place of trees'

When we decided to start the nursery we didn't have any idea what to call it. Then one of our friends said, 'Of course you'll call it

Moormurng Nurseries', and we thought, perfect! We have a state forest on our boundary called Moormurng Forest, and we thought that, if we named the nursery after the forest, people would know where we are.

As it turns out, not many people know about the forest, so we have probably put it on the map. Everybody asks me what 'Moormurng' means, and I used to jokingly say, 'Dead dog in swamp' — I didn't know what it meant. One of my customers looked it up, and she told me its Aboriginal meaning is 'place of trees' or 'bark of trees'. So that was a pleasant surprise.

We started off relatively small-time in 1981. We built the shade house and the hot house, and an area for displays and sales. I'm lucky that my husband is very practical and comes up with labour-saving ideas.

At first we didn't know what to grow or how many. It's a lot of trial and error. Our night reading consisted of plant and propagation books. We also sought out knowledgable people in the area who were very supportive and helpful. You get very knowledgeable yourself about what's indigenous; 'indigenous' was a word that we didn't even know before the nursery. If you're interested in something, the learning comes easily.

'We do it all ourselves'

We have very sandy soil around our house, and it grows most plants beautifully if they get the water to establish themselves. We grow cut-flowers, as well as growing native trees and shrubs in 6-inch tubes, and hybrid shrubs in 3-inch tubes. We also planted proteas and leucadendrons the year the nursery started — once again diversifying and once again not knowing a lot about it.

We have just recently put in nine acres of grape vines, so we are still learning. We are also putting in more Australian foliage and finer flowering plants for mixed bunches. We have extended the acreage around our house to 38, and with a large water-storage unit we are planning on making nearly all of it productive.

Graeme does a lot of the heavy work, and I do the marketing and all of the propagating. He's always in the shed when we do the propagating, filling boxes full of tubes and emptying the trolleys. He very cunningly made all the work benches suitable for my height, so if I do employ somebody, the joke is that they have to be 5'3". Anybody taller gets a sore back.

We did employ people at the start to help with the propagating; a lot of them were friends. Then we had somebody working permanently in the nursery for a twelve-month period, but we realised that we were basically just working to employ someone. Now, with things being a bit tough, we do it all ourselves. Our two boys are older, and they have both served their apprenticeships.

My family were very encouraging from the beginning. My father was really quite thrilled with the idea of the nursery. My father-in-law was not so sure. He put a downer on it initially, but now he sits back and is very pleased for us. But I was never really looking for encouragement; I just did it and it was successful.

'It really is a terrific set-up'

Word of mouth is still our best form of advertising. Even though we've been in the nursery business for sixteen years, I will still have people come and say, 'I didn't know you were here'. I advertise in the local papers regularly, but if they are not interested they don't see it.

We started off selling from home and doing some markets, but markets are a lot of work. Most of the time I could make more money at home in the nursery on a market day than I could at the market. We now support only one field day — the East Gippsland Field Day.

The nursery is open two days a week, Monday and Saturday, and the rest of the week is by appointment. The majority of our customers buy 100 or more trees, so we are mainly a wholesale business. It really is a terrific set-up: I'm not tied to the nursery seven days a week, but during the peak period I have customers most days, at a time that is convenient to us both.

Farmers haven't been our major customers. A lot of the time they don't have the income to buy trees. Trees are a luxury. If you are going to put in a wind-break, you've got to fence, rip and poison. After fencing costs and chemical costs, buying a couple of hundred trees is cheap. It's establishing, planting and maintenance — that's where the money comes into it.

Now with Landcare groups, farmers are getting the opportunity to plant more. Our biggest customers have been hobby farmers wanting to improve and beautify their land, government bodies, and now Landcare groups.

I'd say the majority of my customers do not know what they want. I put them through the third degree. When they say they want some trees, I ask, 'Why? What's the soil type? What look do you want in the future? What are you trying to achieve?'

'I offer a service, too'

I won't sell trees that are going to fail. One day I had a customer who was putting trees on his block but not fencing them off; he was then going to graze sheep. He was a city person, and he proceeded to tell me that sheep eat grass. I said, 'And they will also eat your trees'. I wouldn't sell them to him, so he left. Twelve months later, he came back. He'd built his fences and got himself all organised. He bought his trees, and now he says they are looking great.

So I hope I offer a service, too. I'm happy for people to bring in plans of what they're trying to do. I've got one lady who takes photos. I've never been to her place, but I feel I know it.

If I have a family out and they've got kids that have been really good, I'll give them a tree. I did get caught one day when one little boy said, 'Dad, will I go and get the others?' About eight kids piled out of this van and stood there looking at me. But that was fine. It's nice to give them treats like that. They are all potential customers.

I also get visits from apprentices at Sale TAFE. They might have a couple of days allocated for growing trees, so I'll gladly show them how to do it. I even gave trees away to the last lot — they were such nice kids.

I have observed that only nice people buy trees. Rarely do I get somebody who's not nice. I couldn't have picked a better business. I'm dealing with terrific people 99 per cent of the time.

'It's our major source of income'

I do the books. At the end of the month I take out all the cheque-books and pay the accounts. I'm computer illiterate, and I've been putting off learning to use it because I think, if I do, I'll get stuck with the correspondence and everything else. Graeme's very good with the computer, so when it's time to do the books at the end of the year we both sit down together and work our way through. We vow and declare we'll do it at the end of every month, but it never happens. We're trying. We've got a new program that is so simple that even I can use it. I think I will have to force myself to do it this year.

The income we get from the nursery has made a big difference. In fact, it's our major source of income; the farm has never really been a source of income. We were lucky one year with the wool prices; we actually got a cheque off the farm that year that enabled us to extend our house. We're not complaining. But that was only one year in 23 years. It's the nursery that enables us to live here and continue to farm.

With the drought at the moment, I don't have any sales. We've still got water and I've got beautiful stock, but nobody's going to buy it. It's pretty depressing. And now I've got to do my planning for next year. What do I do? Keep looking ahead to a brighter future. It will rain.

Basically, we only sell between autumn and the end of spring. We don't sell over summer — that's when we do all our propagating. Trees planted during summer grow really well if you water them, but people aren't prepared to put the time in. This year I'll probably carry a lot of my stock over. As long as the water behaves itself I will be able to keep it for next year.

Our biggest expense is the bulk potting-mix, and we buy truck-loads of it. I encourage my customers to bring back the polystyrene boxes and tubes. If they do that I give them some extra trees for their trouble. And I explain to them that if I can get my tubes and boxes back, then I won't be forced to put my prices up. They're happy to bring them back.

It's difficult at the moment with the drought, but it fluctuates. One year you can run out of stock very early, and another year you can have beautiful stock and nobody's interested. The economic climate has definitely affected us. I think 1994 was an exceptionally flat year, but 1996 was very good. And this year *would have* been fairly good, too, if it had been wetter.

'It's a lifestyle'

I really don't know whether I want to be doing this in another ten years time. We're thinking of making the 38 acres here absolutely beautiful for somebody else. With the grape-vines that we've put in and the proteas and the nursery, it would be both a great business and a beautiful place to live. You could also bring people in off the highway very easily. It would be something that maybe a younger couple could do. I know my children aren't interested in doing it themselves.

But, really, it's just a dream; we'll still be here. The drought gets you down. No matter what we do, we seem always to be relying on the weather. We haven't been smart enough to get away from that yet!

But it's so beautiful. What's that saying? 'It's not a living, it's a lifestyle.' And it is true. It's beautiful, and we're very happy.

ENTERPRISE INFORMATION

Name of enterprise
Moormurng Nurseries

Address
PO Box 328
Bairnsdale VIC 3875
Phone/fax (03) 5152 2100

Type of farm
Sheep and wool

Year established
1981

Product
Native trees and shrubs, wind-breaks, ornamentals

District
East Gippsland

Numbers employed
Self, husband and two sons

Market
Farmers, hobby farmers, home gardeners, government

Annual turnover
$20,000-$40,000

Further Information

Please note that the information provided in this section, and in the information boxes placed throughout the text, was correct at the time of the book going to press — as far as we know. We have avoided listing telephone numbers in most cases, as these are often subject to change. If readers notice any errors or omissions, or have suggestions for additional material to be included in future editions, the editor would be happy to hear from you. All correspondence on the subject should be sent to the publishers.

Support Groups

AUSTRALIAN FEDERATION OF BUSINESS AND PROFESSIONAL WOMEN (BPW)

BPW is part of an international network of support and leadership for women in work, education, and personal development. It is a federation of local support groups within which women can build their knowledge, self-confidence, and skills. Career advice and mentorships are offered along with forums, workshops, and discussion groups on such topics as management, finances, technical skills, and other business or social issues.

Membership is by subscription, which then provides a range of opportunities for learning and personal support.

Contact: BPW
PO Box 617
Belconnen ACT 2616

AUSTRALIAN WOMEN IN AGRICULTURE (AWiA)

Incorporated in 1994, this organisation represents the interests of women working in Australian agriculture at the national level. Affiliated with the National Farmers' Federation, AWiA takes a lobbying role in national and state government policies. With branches in each state, the organisation offers opportunities for women to communicate and work together on all aspects of agriculture. AWiA has a regular newsletter, local meetings and gatherings, a web site, and activities such as leadership training, seminars, and product-development groups. AWiA has a strong networking role between individuals and organisations, supporting women in their farming and community endeavours.

Contact: The President
AWiA
Bindi Bindi
WA 6574

THE COUNTRY WOMEN'S ASSOCIATION OF AUSTRALIA (CWA)

The CWA was established in 1928 as an association of women to support and strengthen rural family and community life. A national organisation with state branches, the CWA is also affiliated with the international organisation Associated Country Women of the World. Membership is through subscription, and meetings take place at local, regional, state and national level. The CWA provides services to the community as well as its members, including seminars, public speaking, crafts fairs, and action on social issues such as domestic violence and rural poverty.

There are approximately 45,000 members across Australia. Currently the CWA is seeking ways to adapt to modern demands on rural women and to become involved in matters that will attract younger members.

Contact: The President
CWA
'Glenroy', Wallendbeen NSW 2588

FOUNDATION FOR AUSTRALIAN AGRICULTURAL WOMEN (FAAW)

FAAW was established in 1995 as an outcome of the International Conference for Women in Agriculture held in Melbourne in 1994. It is a national organisation with open membership, with an overall aim of enabling women in agricultural industries and rural communities to reach their full potential. FAAW strengthens women through providing resources to existing organisations and to individuals.

FAAW's long-term goal is to develop a substantial philanthropic fund to give financial backing to farming women's community activities. Activities have included cultural and community leadership training, academic and travelling scholarships, exchange, business-training workshops, family-relief funds, women's broadcasting and newsletters, recruitment of women for boards and industry bodies, and formal and informal contributions to relevant state and national policy-making.

Contact: The Executive Officer
FAAW
GPO Box 1634M
Melbourne VIC 3001

GLOBAL NETWORK FOR RURAL WOMEN

The Global Network is an initiative of Australian rural women to communicate with women working in the world's rural areas, particularly in developing countries. The network raises women's hopes through giving them a voice, corporate strength, and increased agricultural and rural-development knowledge.

The idea originated at the 1994 International Conference for Women in Agriculture, in Melbourne. Delegates went to the NGO Forum on Women in Beijing in 1995 to deliver the recommendations of the conference and consult with rural women from around the world. The network proposal was well received, and now around 3,000 women are registered participants. Regional co-ordinators were recruited to link the network globally.

The network has developed international partnerships, established an internet site with discussion facilities, and produces a newsletter.

Contact: Global Network for Rural Women
GPO Box 1634M
Melbourne VIC 3001

NATIONAL RURAL WOMEN'S UNIT

The Rural Women's Unit was formed in 1995 as part of the Commonwealth Department of Primary Industries and Energy (DPIE). It is responsible for raising the status of rural women by gaining greater recognition of their role in primary industries; by increasing women's participation in policy-making; and by ensuring that issues relevant to women, and their families and communities, are addressed through closer consultation with women.

The unit provides advice to the Minister for Primary Industries and Energy and the department executive. It consults with national and state rural women's organisations, primary industry bodies, and agriculture agencies. It also advises other Commonwealth departments on ways of increasing the number of women on boards and committees relevant to rural communities.

Contact: National Rural Women's Co-ordinator
Rural Women's Unit, DPIE
GPO Box 858
Canberra ACT 2601

RURAL WOMEN'S NETWORKS IN THE STATES

The first rural women's network was established in 1986, in Victoria. There are now networks in six Australian states (five of them are administered through state government departments of agriculture). The networks link individuals and groups, giving rural women an active role in public life. Each network regularly distributes a substantial, free newsletter to tens of thousands of women. Rural women are invited to contribute to the newsletter, covering all issues that affect their lives. Seminars, forums, bursaries, and consultations are additional activities. The networks' directions are guided by reference groups drawn from the community of rural women.

Contact:

The President
Queensland Rural Women's Network Inc
'Cooinda'
Winton QLD 4735

Co-ordinator, Rural Women's Network
Office of Rural Communities
NSW Agriculture
Locked Bag 21
Orange NSW 2800

Co-ordinator, Rural Women's Network
Office of Rural Affairs
Dept of Natural Resources & Environment
PO Box 500
East Melbourne VIC 3002

Project Officer
Tasmanian Women in Agriculture
Dept of Primary Industries & Fisheries
PO Box 46
Kings Meadows TAS 7249

Leader, Rural Affairs Unit
Primary Industries SA
GPO Box 1671
Adelaide SA 5001

Co-ordinator, Rural Women's Network
Dept of Commerce & Trade
PO Box 7234, Cloisters Square
Perth WA 6850

Business Services

GETTING STARTED

Many people have a great idea for a business which never comes to fruition simply because they do not know how to take the next step of looking at the idea's feasibility or making a business plan. While this can be a daunting task, there are people and organisations with whom you can discuss your idea and who will assist you in assessing its viability. Make an appointment with your rural financial counsellor, your economic development officer at your local shire office, or contact Centrelink [telephone 132468] or your local TAFE for the contact number of the managing agent for NEIS — the New Enterprise Incentive Scheme [see below].

Most banks have a rural lending section. Your prospects of securing a loan from them will be greatly enhanced if you've already had a business plan prepared with the help of a professional such as a rural financial counsellor. There are over 90 rural counsellors across Australia. For the name and address of one near you, contact your local department of agriculture or Country Link at 1800 026222.

NEIS— New Enterprise Incentive Scheme

NEIS assists eligible unemployed people to establish viable new businesses. It offers training in small business management, up to 52 weeks of income support, and continuing mentor and business advice. This scheme has been of invaluable assistance in assisting many rural people to establish their own businesses. Contact Centrelink for information on how to contact the managing agent for NEIS in your area.

INFORMATION SERVICES

There are several services available to give assistance to people who have their own business or wish to gain information on setting up their own business. The following organisations would be a good starting point:

AusIndustry Hotline

Ring 132 846 for information, advice and referral. Also can advise on all business assistance programs available from government.

IP Australia

Ring 1300 651010 for advice on patents, trademarks and copyrights.

Auz Trade Hotline

Ring 132 878 for information and advice on exporting and trade.

Business Information Service re Industry, Science and Tourism

Ring 0262 132200 for advice, information, and referral.

Department of Workplace Relations and Small Business

Ring 02 6243 7333 for information on both of these areas.

Country Link

Ring 1800 026222 to reach an invaluable information and referral source set up especially for rural people. Whenever you are not too sure who to contact for advice and assistance this is *definitely* the number to ring. Country Link also distributes *The Rural Book*, which is an up-to-date guide to Commonwealth services and programs. Copies are available on request, free and post-free to country people.

STATE AND TERRITORY GOVERNMENT BUSINESS INFORMATION AND ADVICE DEPARTMENTS

Details specific to each state can be accessed through the *AusINDUSTRY* internet site at http://www.ausindustry.gov.au

NSW:	State and Regional Development
NT:	Department of Asian Relations, Trade and Industry
QLD:	Department of Tourism, Small Business and Industry
VIC:	Department of State Development
SA:	The Business Service
TAS:	Tasmania Development and Resources
WA:	Department of Commerce and Trade

EDUCATION AND TRAINING

National property-management planning program

This farm-business training program is an Australian government initiative delivered through and tailored to the requirements in each state. It encourages forward planning for the farm business, which requires setting long-term goals, in co-operation with all of the family members. *Farm$mart*, the Victorian version of the program, was launched in 1998.

Now managed by farmers themselves at the local level, the program is based on four tailored group workshops on all aspects of farm and family life, encouraging adequate communication skills to enable family business and succession planning into the long-term future. The program offers training in property and environmental management, productivity, and sustainability. A group of farmers can arrange to have the program brought to their district at no cost to themselves.

The following list gives the different names of the program and the telephone contact numbers in each state:

Farming for the Future	New South Wales Telephone: 02 5391 3792
Future Profit	Queensland Telephone: 07 3225 8714
Farm$mart	Victoria Telephone: 03 9207 5555
Better Business	Western Australia Telephone: 08 9386 4404
Farmwi$e	Tasmania Telephone: 03 6336 5256
Property Management Planning Program	South Australia Telephone: 08 8842 3900
Property Management Planning Program	Northern Territory Telephone: 08 8951 8111

National Property Management Planning Program: 08 8842 1103

Farm Bis

The Farm Business Improvement Program [Farm Bis] was announced as part of the federal government's Agriculture Advancing Australia package. This program began operation at the beginning of the 1998 financial year, and focusses on the training needs of those engaged in agriculture or related businesses. The funds for such training are allocated by state rural adjustment authorities. For more information, contact the latter organisation in your state. For names, addresses and telephone numbers of the authority in your state, check your white pages under 'Government Departments' or ring Country Link on 1800 026222.

Courses in Business/Marketing/Horticulture

Most TAFE colleges and agricultural colleges have courses in all these areas, and many have off-campus courses. All these are in the telephone book. For those on the Internet, most have sites which detail courses available. Also of value are Adult Learning Centres. Another great resource is your local Neighbourhood House: often these are located in small communities and will deliver courses on request. Their resources can also be used by those doing off-campus studies. Any of the following state centres will have details of your nearest Neighbourhood House:

Community and Neighbourhood Houses ACT [CANHACT]
Conder Community House
55 Beaumaris Crescent
Conder ACT 2906

NSW Association of Community and Adult Educational Centres [NACEAC]
Vermont
Baraba NSW 2347

Nhulunbuy Community and Neighbourhood Centre
PO Box 6
Nhulunbuy NT 0881

Inala Community House
38 Sittella St
Inala QLD 4077

Community and Neighbourhood Houses and Centres Association [CANH]
96 Rundle St
Kent Town SA 5057

Tasmanian Association of Community Houses [TACH]
25 Circassion St
St Helens TAS 7216

Association of Neighbourhood Houses & Learning Centres [ANHLC]
9th Floor, Carlow House
289 Flinders Lane
Melbourne VIC 3000

Learning Centre Link
GPO 8252
Perth Business Centre
WA 6849

Organic Growers' Organisations

NEW SOUTH WALES

Henry Doubleday Research Association of Australia Inc
816 Comleroy Road
Kurrajong NSW 2758

The Bio-dynamic Farming and Gardening Association of Australia Inc
[Certifying Body]
PO Box 54
Bellingen NSW 2454

Organic Herb Growers of Australia Inc *[Certifying Body]*
PO Box 6171
Sth Lismore NSW 2480

Coffs Regional Organic Growers Organisation
PO Box 363
Coffs Harbour NSW 2450

Hunter Organic Growers Society
PO Box 403
Cessnock NSW 2325

NSW Organic Growers Association
49 Liverpool Road
Heckenberg NSW 2168

Organic Growers Association of NSW
Box 155
Minto NSW 2566

Tweed Richmond Organic Producers Organisation
PO Box 5076
East Lismore NSW 2480

QUEENSLAND

Organic Federation of Australia *[Certifying Body]*
6 Hobson Court
Chapel Hill QLD 4069

Biological Farmers of Australia *[Certifying Body]*
PO Box 3404
Toowoomba Village Fair
Toowoomba QLD 4350

Brisbane Organic Growers Inc
PO Box 236
Lutwyche WLD 4030

SOUTH AUSTRALIA

National Association for Sustainable Agriculture, Australia Ltd *[Certifying Body]*
PO Box 768
Stirling SA 5152

Organic Vignerons Association of Australia Inc *[Certifying Body]*
PO Box 503
Nuriootpa SA 5355

The Soil Association of South Australia Inc
GPO Box 2497
Adelaide SA 5001

TASMANIA

Organic Gardening & Farming Society Tasmania Inc
PO Box 228
Ulverstone TAS 7315

Tasmanian Organic-Dynamic Producers Co-Operative Inc
C/- Post Office Geeveston
TAS 7116

VICTORIA

Organic Retailers & Growers Association of Australia [ORGAA]
PO Box 12852
A'Beckett St Post Office
Melbourne VIC 3000

[ORGAA is a unique nationwide organisation with a membership comprising growers, retailers, and environmentally aware consumers. It was formed as a Victorian association in 1986 in response to a growing concern regarding our food-production systems.]

Biodynamic Agriculture Association
Main Road
Powelltown VIC 3797

East Gippsland Organic Agriculture Association Inc
PO Box 1263
Bairnsdale Vic 3875

WESTERN AUSTRALIA

Organic Growers Association WA (Inc)
PO Box 213
Wembley WA 6014

Notes

Notes

Notes

Notes